Engineering
Design Methods

Engineering Design Methods

NIGEL CROSS

The Open University, Milton Keynes, UK

JOHN WILEY & SONS

Chichester · New York · Brisbane · Toronto · Singapore

Copyright © 1989 by John Wiley & Sons Ltd.

Library of Congress Cataloging in Publication Data
Cross, Nigel.
    Engineering design methods / Nigel Cross.
       p.   cm.
    Bibliography: p.
    Includes index.
    ISBN 0-471-92215-3
    1. Engineering design.   I. Title.
TA174.C76   1989
620'.00425—dc19   88-33368
                       CIP

British Library Cataloguing in Publication Data
Cross, Nigel, 1942–
    Engineering design methods
    1. Engineering. Design. Techniques
    I. Title
    620'.00422

    ISBN 0-471-92215-3

Typeset by Associated Publishing Services Ltd.
Printed and bound in Great Britain by The Anchor Press, Tiptree, Essex

CONTENTS

ACKNOWLEDGEMENTS

Many of the concepts and approaches in this book have been developed in discussion and collaboration with colleagues in the Design Discipline at The Open University. I am grateful also to Beverley Purdy, who has processed almost all of the words.

The author and publisher gratefully acknowledge the following for permission to reproduce figures:

Figure 1: T. A. Thomas, *Technical Illustration*, McGraw-Hill. Figures 2, 11, 12, 13: B. Hawkes & R. Abinett, *The Engineering Design Process*, Longman. Figure 3: J. Fenton, *Vehicle Body Layout and Analysis*, Mechanical Engineering Publications. Figure 4: C. Moore and Van Nostrand Reinhold. Figure 5: A. Howarth. Figure 10: M. J. French, *Conceptual Design for Engineers*, Design Council. Figures 16, 25, 34, 35, 36, 37, 40, 41, 46, 57, 58, 59: G. Pahl & W. Beitz, *Engineering Design*, Design Council/Springer-Verlag. Figures 17, 18: VDI-Verlag. Figure 19: L. March, *The Architecture of Form*, Cambridge University Press. Figures 23, 67: J. C. Jones, *Design Methods*, John Wiley/David Fulton. Figure 26: E. Tjalve, *A Short Course in Industrial Design*, Butterworth. Figures 27, 53: G. Pitts, *Techniques in Engineering Design*, Butterworth. Figures 32, 33: E Krick, *An Introduction to Engineering*, John Wiley. Figure 42: S. Love, *Planning and Creating Successful Engineered Designs*, Advanced Professional Development. Figures 43, 62: U. Pighini, *Design Studies*, Butterworth Scientific. Figure 45: K. W. Norris, *Conference on Design Methods*, Pergamon. Figures 47, 48: V. Hubka, *Principles of Engineering Design*, Butterworth. Figures 49, 50, 63: K. Ehrlenspiel,

ICED 87. Figures 51, 52: M. Tovey/S. Woodward, *Design Studies*, Butterworth Scientific. Figures 60, 61: M. Shahin, *Design Studies*, Butterworth Scientific. Figures 65, 66: Engineering Industry Training Board. Figure 68: A. H. Redford, *Design Studies*, Butterworth Scientific. Figure 69: H. ElMaraghy, *Developments in Assembly Automation*, IFS. Figures 70, 71, 72: The Open University.

INTRODUCTION

This book is intended for use by teachers and students of engineering design, industrial design and industrial engineering. Its main emphasis is on the design of products that have an engineering content, although most of the principles and approaches that it teaches are relevant to the design of all kinds of products and systems. It is essentially concerned with problem formulation, conceptual and embodiment design, rather than the detail design which is the concern of most engineering texts.

The contents of the book divide into three parts. Chapters 1 to 3 offer an overview of the nature of design activity, the design process and design methods in general. Chapter 1 introduces students to the kinds of activities that designers normally undertake, and discusses the particular nature of design problems and the abilities that designers call upon in tackling them. Chapter 2 reviews several of the models of the design process which have been developed in order to help designers structure their approach to designing, and outlines the need for improved procedures. Chapter 3 reviews the field of design methods, describes a range of methods that helps to stimulate creative design thinking and introduces the rational methods which are presented in the next part of the book.

Chapters 4 to 9 constitute a manual of design methods, presented in an independent learning format. The six chapters follow a typical sequence of activities in the design process, providing instruction in the use of appropriate methods within this process. Each chapter presents a separate method, in a standard format of a step-by-step Procedure, a Summary of the steps and a set of practical Examples concluding with a full worked example.

Finally, Chapter 10 outlines a strategic approach to the design process, utilizing the most appropriate combination of creative and rational methods to suit the designer and the design project. Reflecting the approach that is implicit throughout the book, the emphasis is on a flexible design response to problems and on ensuring a successful outcome in terms of good product design.

The book can be most effectively used in conjunction with projects and exercises that require the analysis and clarification of design problems and the generation and evaluation of design solutions. Although intended primarily for students, the book may also be useful to many practising engineers who found design sadly lacking in their own education!

Chapter 1

THE NATURE OF DESIGN

Design Activities

People have always designed things. One of the most basic characteristics of human beings is that they make a wide range of tools and other artefacts to suit their own purposes. As those purposes change, and as people reflect on the currently available artefacts, so refinements are made to the artefacts and completely new kinds of artefacts are conceived and made.

The wish to design things is therefore inherent in human beings, and 'designing' is not something that has always been regarded as needing special abilities. In traditional, craft-based societies 'designing' is not really separate from 'making'; that is to say, there is usually no prior activity of drawing or modelling before the activity of making the artefact. For example, a potter will make a pot by working directly with the clay and without first making any sketches or drawings of the pot.

In modern, industrial societies, however, the activities of designing and of making artefacts are usually quite separate. The process of making something normally cannot start before the process of designing it is complete. In some cases—for example, in the motor-car industry—the period of designing can take several years, whereas the period of making each individual artefact might be measured only in hours.

Perhaps a way towards understanding this modern design activity is to begin at the end—to work backwards from the point where designing is finished and making can start. If making cannot start before designing is finished, then at least it is clear what the design process has to achieve. It has to provide a complete description of the artefact that is to be made. Almost nothing is left to the discretion of those involved in the process of making the artefact—it is specified down to the most detailed dimensions, to the kinds of surface finishes, to the materials, their colours, and so on.

In a sense, perhaps it does not matter how the designer works, so long as he or she produces that final description of the proposed artefact. When a client asks a designer for 'a design', that is what they want—the description. The focus of all design activities is that end-point.

Communication of designs

The most essential design activity, therefore, is the production of a final description of the artefact. This has to be in a form that is understandable to those who will make the artefact. For this reason, the most widely used form of communication is the drawing. For a simple artefact, such as a door handle, one drawing would probably be enough, but for a larger, more complicated artefact, such as a whole building, the number of drawings may well run into hundreds, and for the most complex artefacts, such as chemical process plants, aeroplanes or major bridges, then thousands of drawings may be necessary.

These drawings will range from rather general descriptions—such as plans, elevations and general arrangement drawings—that give an 'overview' of the artefact, to the most specific—such as sections and details—that give precise instructions on how the artefact is to be made. Because they have to communicate precise instructions, with minimal likelihood of misunderstanding, all the drawings are themselves subject to agreed rules, codes and conventions. These codes cover aspects such as how to lay out on one drawing the different views of an artefact relative to each other, how to indicate different kinds of materials and how to specify dimensions. Learning to read and to make these drawings is an important part of design education.

The drawings will often contain annotations of additional information. Dimensions are one such kind of annotation. Written instructions may also be added to the drawings, such as notes on the materials to be used (Figure 1).

Other kinds of specifications besides drawings may also be required. For example, the designer is often required to produce lists of all the separate components and parts that will make up the complete artefact and an accurate count of the numbers of each component to be used (Figure 2). Written specifications of the standards of workmanship or quality of manufacture may also be necessary. Sometimes an artefact is so complex, or so unusual, that the designer

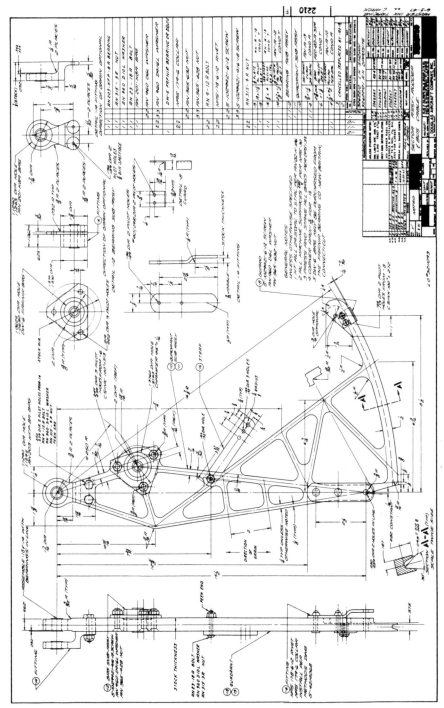

Figure 1. A typical example of a conventional, engineering design detail drawing

makes a complete, three-dimensional mock-up or prototype version in order to communicate the design.

However, there is no doubt that drawings are the most useful form of communication of the description of an artefact that has yet to be made. Drawings are very good at conveying an understanding of what the final artefact has to be like, and that understanding is essential to the person who has to make the artefact.

Nowadays it is not always a *person* who makes the artefact; some artefacts are made by machines that have no direct human operator. These machines might be fairly sophisticated robots or just simple numerically controlled tools such as lathes or milling machines. In these cases, therefore, the final specification of a design prior to manufacture might not be in the form of drawings but in the form of a string of numbers represented on magnetic tape or a computer program that controls the machine's actions.

It is therefore possible to imagine a design process in which no final communication drawings are made. The designer might be able to make a full-size or scale model of the artefact, and measurements could then be taken directly from the model and transmitted to numerically controlled machines.

PARTS LIST

Description	Qty.
Mixer frame	1
Bush	2
Rear drum shield	1
Self-tapping screw	4
Pinion shaft	1
Driven pulley	1
Grub screw	2
Motor pulley	1
Electric motor	1
Motor shaft	1
Hex screw	8
Spring washer	8
Hex nut	8
'Vee' belt	1
Motor cover	1
Tipping handle	1
Plastic handgrip	2
Wheel	2
Starlock washer	2
Mixing drum	1
Drum shaft	1
Bush	2
Thrust washer	1
Drum cap	1
Grommet	1
Allen key	1
Spanner	2
Washer	1
Fitting cord	1
Tubular stand	1
Rubber stop	4

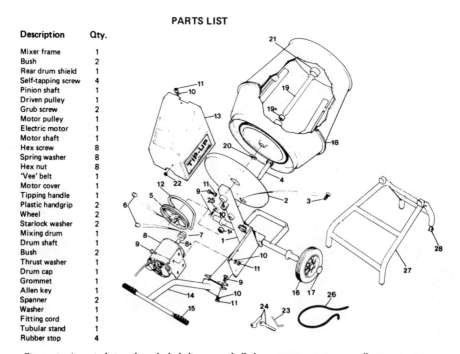

Figure 2. A parts list and exploded diagram of all the components in a small concrete mixer

Evaluation of designs

For the foreseeable future, however, drawings of various kinds will still be used elsewhere in the design process. Even if the final description is to be in the form of a string of numbers, the designer will probably want to make drawings for other purposes.

One of the most important of these other purposes is the checking, or evaluating, of design proposals before deciding on a final version for manufacture. The whole point of having the process of design separated from the process of making is that proposals for new artefacts can be checked before they are put into production. At its simplest, the checking procedure might merely be concerned with, say, ensuring that different components will fit together in the final design; this is an attempt to foresee possible errors and to ensure that the final design is workable. More complicated checking procedures might be concerned with, say, analysing the forces in a proposed design to ensure that each component is designed to withstand the loads on it; this is a process of refining a design to meet certain criteria such as maximum strength, or minimum weight or cost (Figure 3).

This process of refinement can be very complicated and can be the most time-consuming part of the design process. Imagine, for example, the design of a bridge. The designer must first propose the form of the bridge and the materials of which it will be made. In order to check that the bridge is going to be strong enough and stiff enough for the loads that it will carry, the designer must analyse the structure to determine the ways in which loads will be carried by it, what those loads will be in each member of the structure, what deflections will occur, and so on. After a first analysis, the designer might realize—or at least suspect—that changing the locations or angles of some members in the bridge will provide a more efficient distribution of loadings throughout the whole structure. However, these changes will mean that the whole structure will have to be reanalysed and the loads recalculated.

In this kind of situation, it can be easy for the designer to become trapped in an iterative loop of decision-making, where improvements in one part of the design lead to adjustments in another part which lead to problems in yet another part. These problems may mean that the earlier 'improvement' is not feasible. This *iteration* is a common feature of designing.

Nevertheless, despite these potential frustrations this process of refinement is a key part of designing. It consists, firstly, of analysing a proposed design, and for this the designer needs to apply a range of engineering science or other knowledge. In many cases, specialists with more expert knowledge are called in to carry out these analyses. Then, secondly, the results of the analysis are evaluated against the design criteria—does the design come within the cost limit, does it have enough space within it, does it meet the minimum strength requirements, does it use too much fuel, and so on? In some cases, such criteria

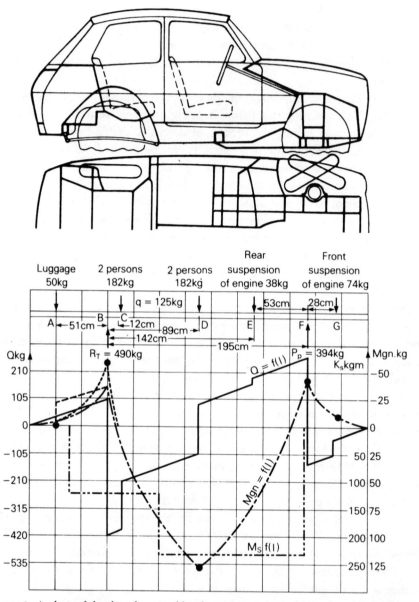

Figure 3. Analysis of the shear forces and bending moments in the body of a small automobile

are set by government regulations or by industry standards; others are set by the client.

Many of the analyses are numerical calculations, and therefore again it is possible to imagine that drawings might not be necessary. However, specialists who are called in to analyse certain aspects of the design will almost certainly want a drawing—or other model of the design—before they can start work.

Visualizations of the proposed design may also be important for the client and designer to evaluate aspects such as appearance, form and colour.

Generation of designs

Before any of these analyses and evaluations can be carried out the designer must, of course, first generate a design proposal. This is often regarded as the mysterious, creative part of designing–the client makes what might well be a very brief statement of requirements and the designer responds (after a suitable period of time) with a design proposal, as if from nowhere. In reality, the process is less 'magical' than it appears.

In most cases, for instance, the designer is asked to design something similar to that which he or she has designed before, and therefore there is a stock of previous design ideas on which to draw. In some cases, only minor modifications are required to a previous design.

Nevertheless, there is something mysterious about the human ability to propose a design for a new—or even just a modified—artefact. It is perhaps as mysterious as the human ability to speak a new sentence—whether it is completely new or just a modification of one heard, read or spoken before.

This ability to design depends partly on being able to visualize something internally, in 'the mind's eye', but perhaps it depends even more on being able to make external visualizations. Once again, drawings are a key feature of the design process. At this early stage of the process, the drawings that the designer makes are not usually meant to be communications to anyone else (Figure 4). Essentially, they are communications with oneself—a kind of thinking aloud.

This early sketching of tentative ideas is necessary because normally there is no way of directly generating an 'optimum' solution from the information provided in the design brief. Quite apart from the fact that the client's brief to the designer may be rather vague, there will be a wide range of criteria to be satisfied, and probably no single objective that must be satisfied above all others (Figure 5).

At the start of the design process, the designer is usually faced with a very poorly defined problem; yet he or she has to come up with a well-defined solution. If one thinks of the problem as a territory, then it is largely unexplored and unmapped, and perhaps imaginary in places! Equally, if one thinks of all potential solutions occupying a kind of solution space, then that, too, is relatively undefined–perhaps infinite. The designer's difficulties are therefore twofold: understanding the problem and finding a solution.

Often these two complementary aspects of design—problem and solution—have to be developed side by side. The designer makes a solution proposal and uses that to help understand what the problem 'really' is and what appropriate solutions might be like. The very first conceptualizations and representations of problem and solution are therefore critical to the kinds of searches and other procedures that will follow, and so to the final solution that will be designed.

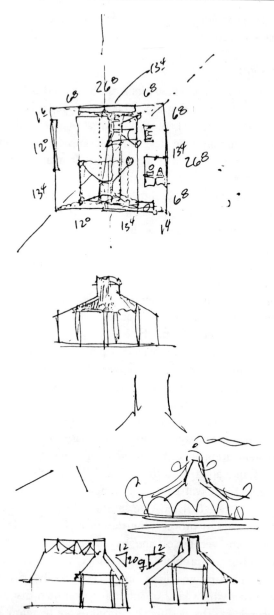

Figure 4. Early sketch design drawings for a small house by the architect Charles Moore

Design Problems

Design problems normally originate as some form of problem statement provided to the designer by someone else—the client or the company management. These problem statements—normally called a design 'brief'—can vary widely in their form and content. At one extreme, they might be something like the statement

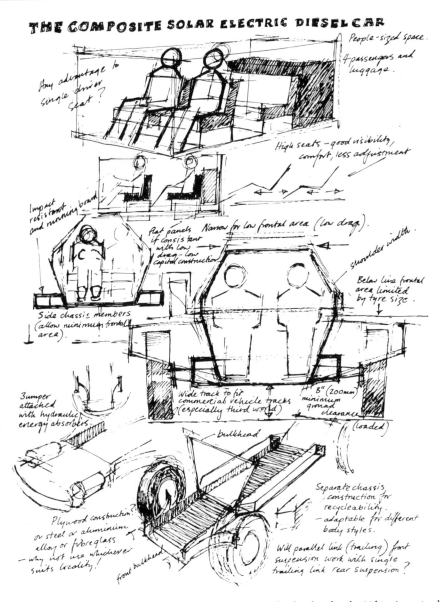

Figure 5. First steps from design brief to design concept: early sketches for the 'Africar'—a simple but robust automobile suitable for Third World conditions

made by President Kennedy in 1961, setting a goal for the United States, 'before the end of the decade, to land a man on the moon and bring him back safely'. In this case, the goal was fixed, but the means of achieving it were quite uncertain. The only constraint in the 'brief' was one of time—'before the end of the decade'. The designers were given a completely novel problem, a fixed goal, only one constraint and huge resources of money, materials and people. This is quite an unusual situation for designers to find themselves in!

At the other extreme is the example of the brief provided to the industrial designer Eric Taylor, for an improved pair of photographic forceps. According to Taylor, the brief originated in a casual conversation with the company's Managing Director, who said to him, 'I was using these forceps last night, Eric. They kept slipping into the tray. I think we could do better than that.' In this case, the brief implied a design modification to an existing product, the goal was rather vague—'forceps that don't slip into the tray'—and the resources available to the designer would have been very limited for such a low-cost product.

Somewhere between these extremes would fall the more normal kind of design brief. A typical example might be the following brief provided to the design department by the planning department of a company manufacturing plumbing fittings. It is for a domestic hot and cold water mixing tap that can be operated with one hand.

One-handed water mixing tap

Required: one-handed household water mixing tap with the following characteristics:

Throughput	10 l/min
Maximum pressure	6 bar
Normal pressure	2 bar
Hot water temperature	60°C
Connector size	10 mm

Attention to be paid to appearance. The firm's trade mark to be prominently displayed. Finished product to be marketed in two year's time. Manufacturing costs not to exceed $20 each at a production rate of 3000 taps per month.

What all of these three examples of design problems have in common is that they set a *goal*, some *constraints* within which the goal must be achieved, and some *criteria* by which a successful solution might be recognized. They do not specify what the solution will be, and there is no certain way of proceeding from the statement of the problem to a statement of the solution—except by 'designing'. Unlike some other kinds of problems, the person setting the problem does not know what 'the answer' is, but they will recognize it when they see it.

Even this last statement is not always true; sometimes clients do not 'recognize' the design solution when they see it. A famous example of early Modern

Architecture is the 'Tugendhat House' in Germany, designed in the 1920s by Mies van der Rohe. Apparently the client had approached the architect after seeing some of the much more conventional houses that he had designed. According to van der Rohe, when he showed the surprising, new design to the client, 'He wasn't very happy at first. But then we smoked some good cigars, . . . and we drank some glasses of good Rhein wine, . . . and then he began to like it very much.'

So the solution that the designer generates may be something that the client 'never imagined might be possible' or perhaps even 'never realized was what they wanted'. Even a fairly precise problem statement gives no indication of what a solution *must* be. It is this uncertainty that makes designing such a challenging activity.

Ill-defined problems

The kinds of problems that designers tackle are regarded as 'ill-defined' or 'ill-structured', in contrast to well-defined or well-structured problems such as chess-playing, crossword puzzles or standard calculations. *Well-defined* problems have a clear goal, often one correct answer and rules or known ways of proceeding that will generate an answer. The characteristics of *ill-defined* problems can be summarized as follows:

1. *There is no definitive formulation of the problem.* When the problem is initially set, the goals are usually vague and many constraints and criteria are unknown. The problem context is often complex and messy, and poorly understood. In the course of problem-solving, temporary formulations of the problem may be fixed, but these are unstable and can change as more information becomes available.

2. *Any problem formulation may embody inconsistencies.* The problem is unlikely to be internally consistent; many conflicts and inconsistencies have to be resolved in the solution. Often, inconsistencies emerge only in the process of problem-solving.

3. *Formulations of the problem are solution-dependent.* Ways of formulating the problem are dependent upon ways of solving it; it is difficult to formulate a problem statement without implicitly or explicitly referring to a solution concept. The way the solution is conceived influences the way the problem is conceived.

4. *Proposing solutions is a means of understanding the problem.* Many assumptions about the problem and specific areas of uncertainty can be exposed only by proposing solution concepts. Many constraints and criteria emerge as a result of evaluating solution proposals.

5. *There is no definitive solution to the problem.* Different solutions can be equally valid responses to the initial problem. There is no objective true-or-false

evaluation of a solution, but solutions are assessed as good or bad, appropriate or inappropriate.

Design problems are widely recognized to be ill-defined problems. It is usually possible to take some steps towards improving the initial definition of the problem, by questioning the client, collecting data, carrying out research, etc. There are also some rational procedures and techniques that can be applied in helping to solve ill-defined problems. However, the designer's traditional approach, as suggested in some of the statements about ill-defined problems listed above, is to try to move fairly quickly to a potential solution, or set of potential solutions, and to use that as a means of further defining and understanding the problem.

Problem structure

However, even when the designer has progressed well into the definition of a solution, difficulties in the problem structure may well still come to light. In particular, sub-solutions can be found to be interconnected with each other in ways that imply a 'pernicious' structure to the problem—e.g. a sub-solution that resolves a particular sub-problem may create irreconcilable conflicts with other sub-problems.

An example of this 'pernicious' problem structure was found in a study of housing design by Luckman (1984). The architects identified five decision areas, or sub-problems, concerned with the directions of span of the roof and first-floor joists, and the provision of load-bearing or non-load-bearing walls and partitions at ground- and first-floor levels. Making a decision in one area (say, the direction of roof span) has implications for the first-floor partitions, and therefore the ground-floor partitions, which has implications for the direction of span of first-floor joists, and therefore for which of the external walls would have to be designed to be load-bearing. This not only had implications for the design of the external wall elevations but also for the direction of span of the roof, and so they came full-circle back to the first decision area. This problem structure is shown diagrammatically in Figure 6, illustrating the cyclical structure that is often found in design problems.

As part of the research study, the individual options in each decision area were separated out and the incompatible pairs of options identified. This is shown in Figure 7, where the links drawn between options indicate that these are incompatible pairs. With this approach, it was possible to identify all the feasible solutions (i.e. sets of five options containing no incompatible pairs). There were found to be eight feasible solutions, and relative costings of each could indicate which would be the cheapest solution.

This example shows that a rigorous approach can sometimes be applied even when the problem appears to be ill-defined and the problem structure pernicious. This lends some support to those who argue that design problems are not always as ill-defined or ill-structured as they might appear to be. In particular, H.A. Simon

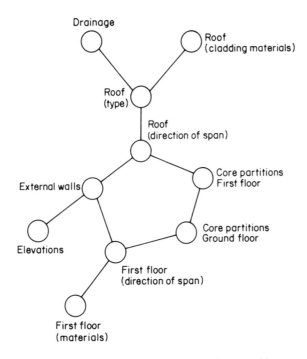

Figure 6. Problem structure in a housing design problem

(1984) has argued that there is no very clear boundary between 'ill-structured' and 'well-structured' problems, and that problems that have been regarded as 'ill-structured' can sometimes, by research and analysis, be reformulated as 'well-structured' problems.

Problem-solving strategies

Research has also shown that designers often attempt to avoid cycling around the pernicious decision loops of design problems by making high-level strategic decisions about solution options. Having identified a number of options, the designer selects what appears to be the best one for investigation at a more detailed level; again, several options are usually evident, and again the best is chosen. This results in what is known as a 'decision tree', with more and more branches opening from each decision point. An example is shown in Figure 8, based on a study by Marples (1960) of a valve design problem for a nuclear reactor. From ten different identified valve arrangements, one was selected for further development. This threw up a number of sub-problems, each of which had alternative solution possibilities; the best was selected, and so on.

This hierarchical 'top-down' approach to design is quite common, although sometimes a 'bottom-up' approach is used, starting with the lowest-level details

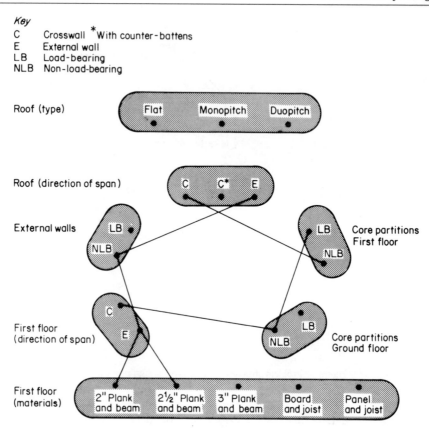

Figure 7. Expanded version of the housing design problem structure, showing incompatible pairs of sub-solutions

and building up to a complete overall solution concept. The decision tree approach perhaps implies that the result is the best possible design, since the best options are chosen at each level. However, a decision at any particular level may well turn out to be sub-optimal in the light of subsequent options available at the other levels. For this reason, there is frequent back-tracking up and down the levels of hierarchy in the design tree.

Design Ability

The world is full of tools, utensils, machines, buildings, furniture, clothes and so many other things that human beings apparently need or want in order to make their lives better. In fact, everything around us that is not a simple piece of Nature has been designed by someone. Even a blank piece of paper has had design decisions made about its size, colour, density, opacity, absorbency, and so on.

Although there is so much design activity going on in the world, the ways in which people design are actually rather poorly understood. Until recently, design

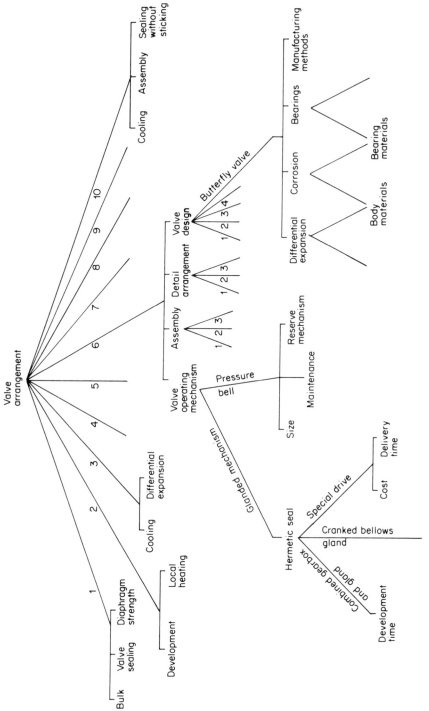

Figure 8. Part of the problem structure in a nuclear reactor valve design problem

ability has been taken for granted as something that many people have and that a few people have a particular talent in. However, there is now a growing body of knowledge about the nature of designing, about design ability and how to develop it, and about the design process and how to improve it.

When designers are asked to discuss their abilities and to explain how they work, a few common themes emerge. One theme is the importance of creativity and 'intuition' in design—even in engineering design. For example, the engineering designer Jack Howe has said:

> I believe in intuition. I think that's the difference between a designer and an engineer. . . . I make a distinction between engineers and engineering designers. . . . An engineering designer is just as creative as any other sort of designer.

Some rather similar comments have been made by the industrial designer Richard Stevens:

> A lot of engineering design is intuitive, based on subjective thinking. But an engineer is unhappy doing this. An engineer wants to test; test and measure. He's been brought up this way and he's unhappy if he can't prove something. Whereas an industrial designer, with his Art School training, is entirely happy making judgements which are intuitive.

Another theme that emerges from designers' own comments is based on the recognition that problems and solutions in design are closely interwoven—that 'the solution' is not always a straightforward answer to 'the problem'. For example, the furniture designer Geoffrey Harcourt commented on one of his creative designs like this:

> As a matter of fact, the solution that I came up with wasn't a solution to the problem at all. I never saw it as that. . . . But when the chair was actually put together (it) in a way quite well solved the problem, but from a completely different angle, a completely different point of view.

A third common theme to emerge is the need to use sketches, drawings or models of various kinds as a way to explore the problem and solution together, and simply to make some progress when faced with the complexity of design. For example, Jack Howe has said that, when uncertain how to proceed:

> I draw *something*. Even if it's 'potty', I draw it. The act of drawing seems to clarify my thoughts.

Given these insights into the nature of design, it is not surprising that the engineering designer Ted Happold should suggest that:

I really have, perhaps, one real talent; that is that I don't mind at all living in the area of total uncertainty.

The quotations above are taken from interviews conducted with a number of successful and eminent designers*. Their comments support some of the hypotheses that have emerged from more objective observational studies of designers at work and other research that has been conducted into the nature of design. Some of this research supports the view that designers have a particular, 'designerly' way of thinking and working.

Design thinking

A research study by Lawson (1984) compared the ways in which designers (in this case architects) and scientists solved the same problem. The scientists tended to use a strategy of systematically exploring the problem, in order to look for underlying rules which would enable them to generate the correct, or optimum, solution. In contrast, the designers tended to suggest a variety of possible solutions until they found one that was good or satisfactory. The evidence from the experiments suggested that scientists problem-solve by analysis, whereas designers problem-solve by synthesis; scientists use 'problem-focused' strategies and designers use 'solution-focused' strategies.

The problem-solving strategies used by designers probably reflect the nature of the problems that they normally tackle. These problems cannot be stated sufficiently explicitly that solutions can be derived directly from them. The designer has to take the initiative in finding a starting point and suggesting tentative solution areas. 'Solution' and 'problem' are then both developed in parallel, sometimes leading to a creative redefinition of the problem or to a solution that lies outside the boundaries of what was assumed to be possible.

The solution-focused strategies of designers are perhaps the best way of tackling design problems, which are by nature ill-defined. In order to cope with the uncertainty of ill-defined problems, the designer has to have the self-confidence to define, redefine and change the problem as given, in the light of solutions that emerge in the very process of designing. People who seek the certainty of structured, well-defined problems will never appreciate the delight of being a designer!

---

*Davies, R. (1985), A psychological enquiry into the origination and implementation of ideas, MSc Thesis, Dept. of Management Sciences, University of Manchester Institute of Science and Technology.

Chapter 2

THE DESIGN PROCESS

Descriptive Models

In recent years there has been quite a number of attempts to provide models of the design process. Some of these models simply *describe* the sequences of activities that typically occur in designing; other models attempt to *prescribe* a better or more appropriate pattern of activities.

Descriptive models of the design process usually emphasize the importance of generating a solution concept early in the process, thus reflecting the 'solution-focused' nature of design thinking. This initial solution 'conjecture' is then subjected to analysis, evaluation, refinement and development. Sometimes, of course, the analysis and evaluation show up fundamental flaws in the initial conjecture and it has to be abandoned, a new concept generated and the cycle started again. The process is *heuristic*: using previous experience, general guidelines and 'rules of thumb' that lead in what the designer hopes to be the right direction, but with no absolute guarantee of success.

In Chapter 1, I developed a simple descriptive model of the design process, based on the essential activities that the designer performs. The end-point of the process is the *communication* of a design, ready for manufacture. Prior to this, the design proposal is subject to *evaluation* against the goals, constraints and criteria of the design brief. The proposal itself arises from the *generation* of a concept by

19

the designer. Putting these three activity types in their natural sequence, we have a three-stage model of the design process consisting of generation–evaluation–communication.

This simple three-stage model is shown diagrammatically in Figure 9. Assuming that the evaluation stage does not always lead directly onto the communication of a final design, but that sometimes a new, more satisfactory concept has to be chosen, a feedback loop is shown from the evaluation stage to the generation stage.

Models of the design process are often drawn in this flow diagram form, with the development of the design proceeding from one stage to the next, but with feedback loops showing the iterative returns to earlier stages which are frequently necessary. For example, French (1985) has developed a more detailed model, shown in Figure 10, based on the following activities of design:

Analysis of problem
Conceptual design
Embodiment of schemes
Detailing

In the diagram, the circles represent stages reached, or outputs, and the rectangles represent activities, or work in progress.

The process begins with an initial statement of a 'need', and the first design activity is 'analysis of the problem'. French suggests that:

The analysis of the problem is a small but important part of the overall process. The output is a statement of the problem, and this can have three elements:
1. A statement of the design problem proper
2. Limitations placed upon the solution, e.g. codes of practice, statutory requirements, customers' standards, date of completion, etc.
3. The criterion of excellence to be worked to

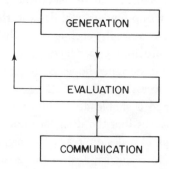

Figure 9. A simple three-stage model of the design process

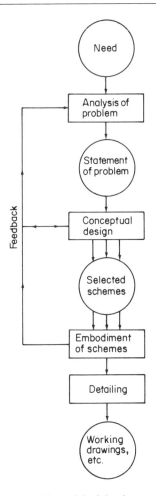

Figure 10. French's model of the design process

These three elements correspond to the goals, constraints and criteria of the design brief.

The activities that follow, according to French, are then:

1. *Conceptual design.* This phase . . . takes the statement of the problem and generates broad solutions to it in the form of schemes. It is the phase that makes the greatest demands on the designer, and where there is the most scope for striking improvements. It is the phase where engineering science, practical knowledge, production methods and commercial aspects need to be brought together, and where the most important decisions are taken.

2. *Embodiment of schemes.* In this phase the schemes are worked up in greater detail and, if there is more than one, a final choice between them is made. The end product is usually a set of general arrangement drawings. There is (or

should be) a great deal of feedback from this phase to the conceptual design phase.

3. *Detailing*. This is the last phase, in which a very large number of small but essential points remains to be decided. The quality of this work must be good, otherwise delay and expense or even failure will be incurred; computers are already reducing the drudgery of this skilled and patient work and reducing the chance of errors, and will do so increasingly.

These activities are typical of conventional engineering design. Figures 11, 12 and 13 illustrate the type of work that goes on in each stage. The illustrations are examples from the design of the small concrete mixer, for which the final design was shown in Figure 2. *Conceptual design* is shown in Figure 11, where three alternatives are proposed for the drive connection from the motor to the mixing drum. *Embodiment design* is shown in Figure 12, where concept (c) is developed in terms of how to support and assemble the motor, drum, pulleys, etc. Figure 13 shows a small example of *detail design*, in which the motor-mounting plate is redesigned from a welded T-shape to a channel section U-shape, after tests of a prototype found excessive vibration occurring in the original.

Prescriptive Models

As well as models that simply describe a more-or-less conventional, heuristic process of design, there have been several attempts at building prescriptive models of the design process. These latter models are concerned with trying to persuade or encourage designers to adopt improved ways of working. They usually offer a

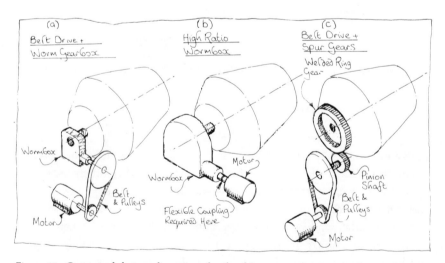

Figure 11. Conceptual design: alternatives for the drive connection in a small concrete mixer

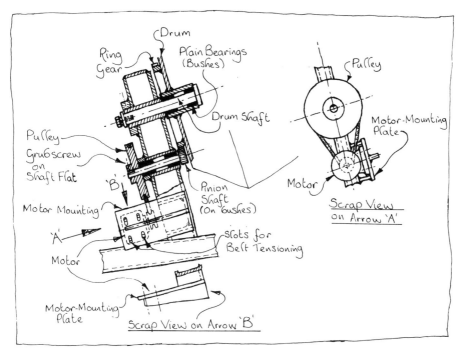

Figure 12. Embodiment design: one concept developed in more detail

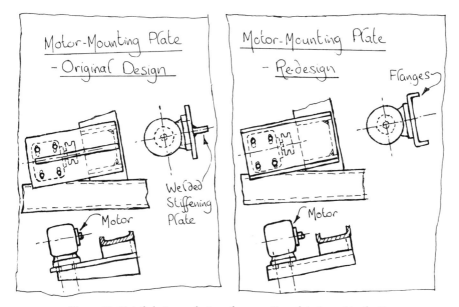

Figure 13. Detail design: redesign of a mounting plate to resist vibration

more *algorithmic*, systematic procedure to follow, and are often regarded as providing a particular *design methodology*.

Many of these prescriptive models have emphasized the need for more analytical work to precede the generation of solution concepts. The intention is to try to ensure that the design problem is fully understood, that no important elements of it are overlooked and that the 'real' problem is identified. There are plenty of examples of excellent solutions to the wrong problem!

These models have therefore tended to suggest a basic structure to the design process of analysis–synthesis–evaluation. These stages were defined by J. C. Jones (1984) in an early example of a systematic design methodology as follows:

1. *Analysis*. Listing of all design requirements and the reduction of these to a complete set of logically related performance specifications.
2. *Synthesis*. Finding possible solutions for each individual performance specification and building up complete designs from these with least possible compromise.
3. *Evaluation*. Evaluating the accuracy with which alternative designs fulfil performance requirements for operation, manufacture and sales *before* the final design is selected.

This may sound very similar to a conventional design process, but the emphases here are on performance specifications logically derived from the design problem, generating several alternative design concepts by building up the best sub-solutions and making a rational choice of the best of the alternative designs. Such sensible, rational procedures are not always followed in conventional design practice!

A more detailed prescriptive model was developed by Archer (1984), and is summarized in Figure 14. This includes interactions with the world outside of the design process itself, such as inputs from the client, the designer's training and experience, other sources of information, etc. The output is, of course, the communication of a specific solution. These various inputs and outputs are shown as external to the design process in the flow diagram, which also features many feedback loops.

Within the design process, Archer identified six types of activity:

1. *Programming*. Establish crucial issues; propose a course of action.
2. *Data collection*. Collect, classify and store data.
3. *Analysis*. Identify sub-problems; prepare performance (or design) specifications; reappraise proposed programme and estimate.
4. *Synthesis*. Prepare outline design proposals.
5. *Development*. Develop prototype design(s); prepare and execute validation studies.
6. *Communication*. Prepare manufacturing documentation.

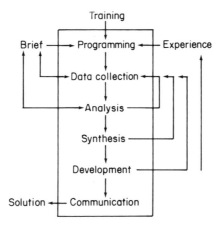

Figure 14. Archer's model of the design process

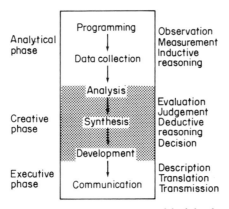

Figure 15. Archer's three-phase summary model of the design process

Archer summarized this process as dividing into three broad phases: analytical, creative and executive (Figure 15). He suggested that:

One of the special features of the process of designing is that the analytical phase with which it begins requires objective observation and inductive reasoning, while the creative phase at the heart of it requires involvement, subjective judgement, and deductive reasoning. Once the crucial decisions are made, the design process continues with the execution of working drawings, schedules, etc., again in an objective and descriptive mood. The design process is thus a creative sandwich. The bread of objective and systematic analysis may be thick or thin, but the creative act is always there in the middle.

Some much more complex models have been proposed, but they often tend to obscure the general structure of the design process by swamping it in the fine detail of the numerous tasks and activities that are necessary in all practical design work. A reasonably comprehensive model that still retains some clarity is that offered by Pahl and Beitz (1984) (Figure 16). It is based on the following design stages:

1. *Clarification of the task.* Collect information about the requirements to be embodied in the solution and also about the constraints.
2. *Conceptual design.* Establish function structures; search for suitable solution principles; combine into concept variants.
3. *Embodiment design.* Starting from the concept, the designer determines the layout and forms and develops a technical product or system in accordance with technical and economic considerations.
4. *Detail design.* Arrangement, form, dimensions and surface properties of all the individual parts finally laid down; materials specified; technical and economic feasibility re-checked; all drawings and other production documents produced.

Considerable work on these kinds of models and on other aspects of rationalizing the design process has been done in Germany. The professional engineers' body, Verein Deutscher Ingenieure (VDI), has produced a number of 'VDI Guidelines' in this area, including VDI 2221 'Systematic Approach to the Design of Technical Systems and Products'. This Guideline suggests a systematic approach in which 'The design process, as part of product creation, is subdivided into general working stages, making the design approach transparent, rational and independent of a specific branch of industry'.

The structure of this general approach to design is shown in Figure 17, and is based on seven stages, each with a particular output. The output from the first stage, the specification, is regarded as particularly important and is constantly reviewed and kept up-to-date, and is used as a reference in all the subsequent stages.

The second stage of the process consists of determining the required functions of the design and producing a diagrammatic function structure. In stage 3 a search is made for solution principles for all sub-functions, and these are combined in accordance with the overall function structure into a principal solution. This is divided, in stage 4, into realizable modules and a module structure representing the breakdown of the solution into fundamental assemblies. Key modules are developed in stage 5 into a set of preliminary layouts. These are refined and developed in stage 6 into a definitive layout, and the final product documents are produced in stage 7.

In the Guideline it is emphasized that several solution variants should be analysed and evaluated at each stage and that there is a lot more detail in each stage than is shown in the diagram. The following words of warning about the approach are also given:

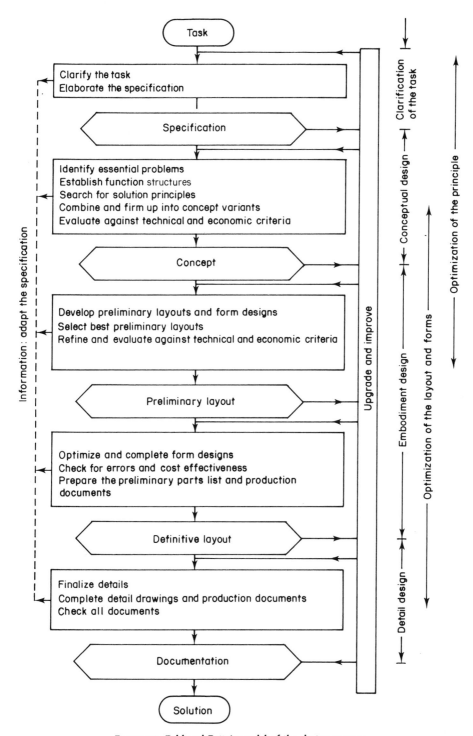

Figure 16. Pahl and Beitz's model of the design process

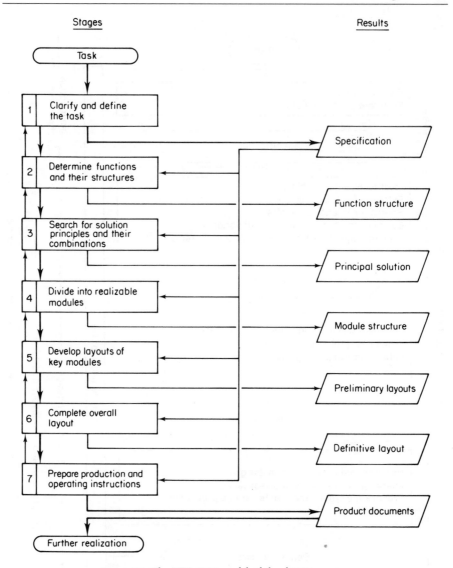

Figure 17. The VDI 2221 model of the design process

It is important to note that the stages do not necessarily follow rigidly one after the other. They are often carried out iteratively, returning to preceding ones, thus achieving a step-by-step optimisation.

The VDI Guideline follows a general systematic procedure of first analysing and understanding the problem as fully as possible, then breaking this into sub-problems, finding suitable sub-solutions and combining these into an overall solution. The procedure is shown diagrammatically in Figure 18.

This kind of procedure has been criticized in the design world because it seems to be based on a problem-focused rather than a solution-focused approach. It therefore runs counter to the designer's traditional ways of thinking.

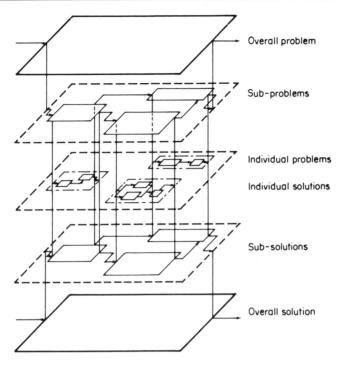

Overall problem

Sub-problems

Individual problems

Individual solutions

Sub-solutions

Overall solution

Figure 18. The VDI 2221 model of development from problem to solution

A more radical model of the design process, which recognizes the solution–focused nature of design thinking, has been suggested by March (1984) (Figure 19). He argued that the two conventionally understood forms of reasoning—inductive and deductive—only apply logically to the evaluative and analytical types of activity in design. However, the type of activity that is most particularly associated with design is that of synthesis, for which there is no commonly acknowledged form of reasoning. March drew on the work of the philosopher C.S. Peirce to identify this missing concept of 'abductive' reasoning. According to Pierce,

> Deduction proves that something *must* be; induction shows that something *actually* is operative; abduction merely suggests that something *may* be.

It is this hypothesizing of what *may* be, the act of synthesis, that is central to design. Because it is the kind of thinking by which designs are generated or produced, March prefers to call it 'productive' reasoning. Thus his model for a rational design process is a 'PDI model'—production–deduction–induction.

In this model, the first phase, of *productive* reasoning, draws on a preliminary statement of requirements and some presuppositions about solution types in order

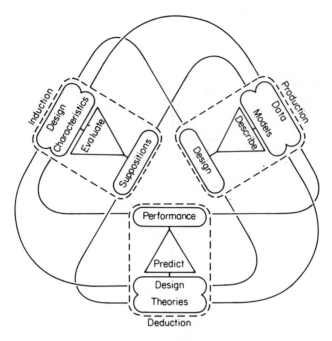

Figure 19. March's model of the design process

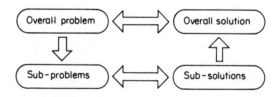

Figure 20. The symmetrical relationships of problem/sub-problems/sub-solutions/solution

to produce, or describe a design proposal. From this proposal and established theory (e.g. engineering science) it is possible *deductively* to analyse, or predict, the performance of the design. From these predicted performance characteristics it is possible *inductively* to evaluate further suppositions or possibilities, leading to changes or refinements in the design proposal.

Certainly it seems that, in most design situations, it is not possible, or relevant, to attempt to analyse 'the problem' *ab initio* and in some form of abstract isolation from solution concepts. Although there may be a logical progression from problem to sub-problems and from sub-solutions to solution, there is a symmetrical relationship between problem and solution and between sub-problems and sub-solutions, as illustrated in Figure 20. The general direction of movement in this model is anticlockwise, but there are substantial periods of toing-and-froing between problem and solution.

Systematic Procedures

There may be differences in their preferred models, but the proponents of systematic procedures all agree that there is an urgent need to improve on traditional ways of working in design.

There are several reasons for this concern to develop new design procedures. One is the increasing complexity of modern design. A great variety of new demands is increasingly being made on the designer, such as the new materials and devices (e.g. electronics) that become available and the new problems that are presented to designers. Many of the products and machines to be designed today have never existed before, and so the designer's previous experience may well be irrelevant and inadequate for these tasks. Therefore a new, systematic approach is needed, it is argued.

A related part of the complexity of modern design is the need to develop team work, with many specialists collaborating in and contributing to the design. To help coordinate the team, it is necessary to have a clear, systematic approach to design, so that specialists' contributions are made at the right point in the process. Dividing the overall problem into sub-problems in a systematic procedure also means that the design work itself can be subdivided and allocated to appropriate team members.

As well as being more complex, modern design work often has very high risks and costs associated with. For example, many products are designed for mass manufacture, and the costs of setting up the manufacturing plant, buying-in raw materials, and so on, are so high that the designer cannot afford to make mistakes: the design must be absolutely right before it goes into production. This means that any new product must have been through a careful process of design. Other kinds of large, one-off designs, such as chemical process plants or complex products such as aeroplanes, also have to have a very rigorous design process to try to ensure their safe operation and avoid the catastrophic consequences of failure.

Finally, there is a more general concern with trying to improve the efficiency of the design process. In some industries there is a pressing need to ensure that the lead-time necessary to design a new product is kept to a minimum. In all cases, it is desirable to try to avoid the mistakes and delays that often occur in conventional design procedures. The introduction of computers already offers one way of improving the efficiency of the design process, and is also in itself an influence towards more systematic ways of working.

Chapter 3

DESIGN METHODS

What are Design Methods?

In a sense, any identifiable way of working, within the context of designing, can be considered to be a design method. The most common design method can be called the method of 'design by drawing'; that is to say, most designers rely extensively on drawing as their main aid to designing.

Design methods are any procedures, techniques, aids or 'tools' for designing. They represent a number of distinct kinds of activities that the designer might use and combine into an overall design process.

Although some design methods can be the conventional, normal procedures of design, such as drawing, there has been a growth in recent years of new, unconventional procedures that are more usually grouped together under the name of 'design methods'. The origins of these new methods go back to the Second World War and the development of techniques for solving the many novel and very pressing problems of warfare. Operational research (OR) originated and developed in this way, and in the nineteen-fifties people began to apply OR and other new techniques in areas such as management and planning. Then, in the nineteen-sixties, designers also began to adapt and develop some of these new techniques, and to invent similar ones of their own, for use in the design process.

The main intention of these new methods is that they attempt to bring logical procedures into the design process. Some of these methods are themselves new inventions, some are adapted from OR, management sciences or other sources, and some are simply extensions or formalizations of the informal techniques that designers have always used. For example, the informal methods of looking up manufacturers' catalogues or seeking advice from colleagues might be formalized into an 'information search' method, or informal procedures for saving costs by detailed redesigning of a component can be formalized into a 'value analysis' method.

It seems that some of the new methods can become *over*-formalized or can be merely fancy names for old, common-sense techniques. They can also appear to be *too* systematic to be useful in the rather messy and often hurried world of the design office. For these kinds of reasons, many designers are still mistrustful of the whole idea of 'design methods'.

The counter-arguments to that view are based on the reasons for adopting systematic procedures which were outlined in Chapter 2. For instance, many modern design projects are too complex to be resolved satisfactorily by the old, conventional methods. There are also too many errors made with conventional ways of working, and they are not very useful where team work is necessary. Design methods try to overcome these kinds of problems and—above all—they try to ensure that a better product results from the new design process.

What kinds of design methods are there?

There have been many new methods developed to help overcome the difficulties of modern design problems. For example, the textbook of *Design Methods* by J.C. Jones (1981) contains descriptions of thirty-five methods, including the following:

Methods of exploring design situations

Method	Aim
Stating objectives	To identify external conditions with which the design must be compatible
Literature searching	To find published information that can favourably influence the designers' output and that can be obtained without unacceptable cost and delay
Searching for visual inconsistencies	To find directions in which to search for design improvements
Interviewing users	To elicit information that is known only to users of the product or system in question
Questionnaires	To collect usable information from the members of a large population

Investigating user behaviour	To explore the behaviour patterns and to predict the performance limits of potential users of a new design
Systemic testing	To identify actions that are capable of bringing about desired changes in situations that are too complicated to understand
Selecting scales of measurement	To relate measurements and calculations to the uncertainties of observation, to the costs of data collecting and to the objectives of the design project
Data logging and data reduction	To infer, and to make visible, patterns of behaviour upon which critical design decisions depend

Methods of searching for ideas

Method	Aim
Brainstorming	To stimulate a group of people to produce many ideas quickly
Synectics	To direct the spontaneous activity of the brain and the nervous system towards the exploration and transformation of design problems
Removing mental blocks	To find new directions of search when the apparent search space has yielded no wholly acceptable solution
Morphological charts	To widen the area of search for solutions to a design problem

Methods of exploring problem structure

Method	Aim
Interaction matrix	To permit a systematic search for connections between elements within a problem
Interaction net	To display the pattern of connections between elements within a design problem
Analysis of interconnected decision areas	To identify and to evaluate all the compatible sets of sub-solutions to a design problem
System transformation	To find ways of transforming an unsatisfactory system so as to remove its inherent faults
Innovation by boundary shifting	To shift the boundary of an unsolved design problem so that outside resources can be used to solve it
Functional innovation	To find a radically new design capable of creating new patterns of behaviour and demand

Alexander's method of determining components	To find the right physical components of a physical structure such that each component can be altered independently to suit future changes in the environment
Classification of design information	To split a design problem into manageable parts

Methods of evaluation

Method	Aim
Checklists	To enable designers to use knowledge of requirements that have been found to be relevant in similar situations
Selecting criteria	To decide how an acceptable design is to be recognized
Ranking and weighting	To compare a set of alternative designs using a common scale of measurement
Specification writing	To describe an acceptable outcome for designing that has yet to be done
Quirk's reliability index	To enable inexperienced designers to identify unreliable components without testing

As can be seen from the list, some methods are formal versions of conventional procedures (e.g. literature searching, interviewing users), some are applications of methods first developed in other fields (e.g. data logging from science; brainstorming from advertising) and some are new inventions (e.g. Alexander's method, Quirk's reliability index). The list also shows that different design methods have different purposes and are relevant to different aspects of and stages in the design process.

What do different design methods have in common?

Looking over all the methods in the previous list, two principal common features emerge. One is that design methods *formalize* certain procedures of design and the other is that design methods *externalize* design thinking. Formalization is a common feature of design methods because it attempts to avoid the occurrence of oversights, of overlooked factors in the design problems and of the kinds of errors that occur with informal methods. The process of formalizing a procedure also tends to widen the approach that is taken to a design problem and to widen the search for appropriate solutions—it encourages and enables you to think beyond the first solution that comes into your head.

This is also related to the other general aspect of design methods—that they externalize design thinking; i.e. they try to get your thoughts and thinking processes out of your head and into the charts and diagrams that commonly feature in design methods. This externalizing is a significant aid when dealing with complex problems, but it is also a necessary part of team work; i.e. providing means by which all the members of the team can see what is going on and can contribute to the design process. Getting a lot of systematic work out of your head and onto paper also means that your mind can be more free to pursue the kind of thinking it is best at—intuitive and imaginative thinking.

Design methods are therefore *not* the enemy of creativity, imagination and intuition. Quite the contrary: they are perhaps more likely to lead to novel design solutions than the informal, internal and often incoherent thinking procedures of the conventional design process. Some design methods are, indeed, techniques specifically for aiding creative thought. In fact, the general body of design methods can be classified into two broad groups: creative methods and rational methods.

Creative Methods

There are several design methods that are intended to help stimulate creative thinking. In general, they work by trying to increase the flow of ideas, by removing the mental blocks that inhibit creativity or by widening the area in which a search for solutions is made.

BRAINSTORMING

The most widely known creative method is brainstorming. This is a method for generating a large number of ideas, most of which will subsequently be discarded, but with perhaps a few novel ideas being identified as worth following up. It is normally conducted as a small group session of about 5–12 people.

The group of people selected for a brainstorming session should be diverse. It should not just be experts or those knowledgeable in the problem area, but should include a wide range of expertise and even laypeople if they have some familiarity with the problem area. The group must be non-hierarchical, although one person does need to take an organizational lead.

The role of the group leader in a brainstorming session is to ensure that the format of the method is followed, and that it does not degenerate into a round-table discussion. An important prior task for the leader is to formulate the problem statement used as a starting point. If the problem is stated too narrowly, then the range of ideas from the session may be rather limited. On the other hand, a very vague problem statement leads to equally vague ideas, which may be of no practical use. The problem can often be usefully formulated as a question, such as 'How can we improve on X?'.

In response to the initial problem statement, the group members are asked to spend a few minutes—in silence—writing down the first ideas that come into their heads. It is a good idea if each member has a pile of small record cards on which to write these and subsequent ideas. The ideas should be expressed succinctly, and written one per card.

The next, and major, part of the session is for each member of the group, in turn, to read out one idea from his or her set. The most important rule here is that *no criticism is allowed* from any other member of the group. The usual responses to unconventional ideas, such as 'That's silly' or 'That will never work', kill off spontaneity and creativity. At this stage, the feasibility or otherwise of any idea is not important—evaluation and selection will come later.

What each group member should do in response to every other person's idea is to try to build on it, take it a stage further, to use it as a stimulus for other ideas or to combine it with his or her own ideas. For this reason, there should be a short pause after each idea is read out, to allow a moment for reflection and for writing down further new ideas. However, the session must not become too stilted; the atmosphere should be relaxed and free-wheeling. A brainstorming session should also be fun: humour is often an essential ingredient of creativity.

The group session should not last more than about 20–30 minutes, or should be wound up when no more new ideas are forthcoming. The group leader, or someone else, then collects all the cards and spends a separate period evaluating the ideas. A useful aid to this evaluation is to sort or classify the ideas into related groups; this in itself often suggests further ideas or indicates the major types of idea that there appear to be. If principal solution areas and one or two novel ideas result from a brainstorming session then it will have been worth while.

Participating in a brainstorming session is rather like playing a party game, and like a party game it only works well when everyone sticks to the rules. In fact, all design methods only work best when they are followed with some rigour, and not in a sloppy or half-hearted fashion. The essential rules of brainstorming are:

No criticism is allowed during the session.
A large quantity of ideas is wanted.
Seemingly crazy ideas are quite welcome.
Keep all ideas short and snappy.
Try to combine and improve on the ideas of others.

SYNECTICS

Creative thinking often draws on analogical thinking—on the ability to see parallels or connections between apparently dissimilar topics. The role of humour is again relevant, since most jokes depend for their effect on the unexpected transfer or juxtaposition of concepts from one context to another, or what Koestler called the 'bisociation' of ideas. Bisociation plays a fundamental role in creativity.

The use of analogical thinking has been formalized in a creative design method known as 'synectics'. Like brainstorming, synectics is a group activity in which criticism is ruled out, and the group members attempt to build, combine and develop ideas towards a creative solution to the set problem. Synectics is different from brainstorming in that the group tries to work collectively towards a particular solution, rather than generating a large number of ideas. A synectics session is much longer than brainstorming and much more demanding. In a synectics session, the group is encouraged to use particular types of anology. These have been summarized by J.C. Jones as follows:

1. *Direct analogies.* These are most readily found by seeking a biological solution to a similar problem, e.g. Brunel's observation of a shipworm forming a tube for itself as it bored through timber. This is said to have led him to the idea of a caisson for underwater constructions.
2. *Personal analogies.* The designer imagines what it would be like to use one's body to produce the effect that is being sought, e.g. what would it feel like to be a helicopter blade, what forces would act on me from the air and from the hub; what would it feel like to be a bed?
3. *Symbolic analogies.* These are poetic metaphors and similies in which aspects of one thing are identified with aspects of another, e.g. the *mouth* of a river, the *head* of a hammer, a *tree* of decisions, *sheet* lightening, to *damp* an oscillation, to *drive* a bargain.
4. *Fantasy analogies.* To wish for, or to imagine, things as they are known not to be, e.g. what we really want is a little slave to dial the telephone for us: we need a road that disappears except where the wheels touch the ground.

A synectics session starts with 'the problem as given'—the problem statement as presented by the client or company management. Analogies are then sought that help to 'make the strange familiar', i.e. expressing the problem in terms of some more familiar (but perhaps rather distant) analogy. This leads to a conceptualization of 'the problem as understood'–the key factor or elements of the problem that need to be resolved, or perhaps a complete reformulation of the problem. The problem as understood is then used to guide the use of analogies again, but this time to 'make the familiar strange'. Unusual, creative analogies are sought, which may lead to novel solution concepts. The analogies are used to open up lines of development which are pursued as hard and as imaginatively as possible by the group.

ENLARGING THE SEARCH SPACE

A common form of mental block to creative thinking is to assume rather narrow boundaries within which a solution is sought. Many creativity techniques are aids to enlarging the 'search space'.

Transformation One such technique attempts to 'transform' the search for a solution from one area to another. This often involves applying verbs that will transform the problem in some way, such as

> magnify, minify, modify, unify, subdue, subtract, add, divide, multiply, repeat, replace, relax, dissolve, thicken, soften, harden, roughen, flatten, rotate, rearrange, reverse, combine, separate, substitute, eliminate.

Random input Creativity can be triggered by random inputs from whatever source. This can be applied as a deliberate technique, e.g. opening a dictionary or other book and choosing a word at random and using that to stimulate thought on the problem in hand.

Why? Why? Why? Another way of extending the search space is to ask a string of questions 'why?' about the problem, such as 'why is this device necessary?' 'Why can't it be eliminated?', etc. Each answer is followed up, like a persistent child, with another 'Why?' until a dead end is reached or an unexpected answer prompts an idea for a solution. There may be several answers to any particular 'Why?', and these can be charted as a network of question-and-answer chains.

Counter-planning This method is based on the concept of the dialectic, i.e. pitting an idea (the thesis) against its opposite (the antithesis) in order to generate a new idea (the synthesis). It can be used to challenge a conventional solution to a problem by proposing its deliberate opposite and seeking a compromise. Alternatively, two completely different solutions can be deliberately generated, with the intention of combining the best features of each into a new synthesis.

The creative process

The methods above are a few techniques that have been found useful when it is necessary for a designer or design team to 'turn on' their creative thinking. However, creative, original ideas can also seem to occur quite spontaneously, without the use of any such aids to creative thinking. Is there, therefore, a more general process of creative thinking that can be developed?

Psychologists have studied accounts of creative thinking from a wide range of scientists, artists and designers. In fact, as most people have also experienced, these highly creative individuals generally report that they experience a very sudden creative insight that suggests a solution to the problem they have been working on. There is a sudden 'illumination'—just like the light-bulb flashing on that cartoonists use to suggest someone having a bright idea.

This creative 'Ah-ha!' experience often occurs when the individual is not expecting it, and after a period when they have been thinking about something else. This is rather like the common phenomenon of suddenly remembering a name or word that could not be recalled when it was wanted.

However, the sudden illumination of a bright idea does not usually occur without considerable background work on a problem. The illumination or key insight is also usually just the germ of an idea that needs a lot of further work to develop it into a proper, complete solution to the problem.

Similar kinds of thought sequence occur often enough in creative thinking for the psychologists to suggest that there is a general pattern to it. This general pattern is the sequence:

Recognition–preparation–incubation–illumination–verification

1. *Recognition* is the first realization or acknowledgement that a 'problem' exists.
2. *Preparation* is the application of deliberate effort to understand the problem.
3. *Incubation* is a period of leaving it to 'mull over' in the mind, allowing one's subconscious to go to work.
4. *Illumination* is the (often quite sudden) perception or formulation of the key idea.
5. *Verification* is the hard work of developing and testing the idea.

This process is essentially one of work–relaxation–work, with (if you are lucky) the creative insight occurring in a relaxation period. The hard work of preparation and verification is essential. Like most other kinds of creative activity, creative design is one per cent inspiration and ninety-nine per cent perspiration!

Rational Methods

More commonly regarded as 'design methods' than the creativity techniques are the rational methods that encourage a systematic approach to design. Nevertheless, these rational methods often have similar aims to the creative methods, such as widening the search space for potential solutions or facilitating team work and group decision-making. So it is not necessarily true that rational methods are somehow the very opposite of creative methods.

Many designers are suspicious of rational methods, fearing that they are a 'straightjacket' or that they are stifling to creativity. This is a misunderstanding of the intentions of systematic design, which is meant to improve the quality of design decisions and hence of the end product. Creative methods and rational methods are complementary aspects of a systematic approach to design. Rather than a 'straightjacket', they should be seen as 'lifejacket', helping the designer to keep afloat.

Perhaps the simplest kind of rational method is the checklist. Everyone uses this method in daily life—for example, in the form of a shopping list or list of things to remember to do. It *externalizes* what you have to do, so that you do not have to try to keep it all in your head and so that you do not overlook something. It *formalizes* the process, by making a record of items which can be checked off as they are collected or achieved until everything is complete. It also allows team work or participation by a wider group—e.g. all the family can contribute suggestions for the shopping list. It also allows subdivision of the task (i.e. improving the efficiency of the process), such as allocating separate sections of the list to different members of the team. In these respects, it is a model for most of the rational design methods. In design terms, a checklist may be a list of questions to be asked in the initial stages of design, a list of features to be incorporated in the design, or a list of criteria, standards, etc, that the final design must meet.

There is a wide range of rational design methods, as we saw in the long list above, covering all aspects of the design process from problem clarification to detail design. The next six chapters present a selection of the most relevant and widely used methods, also covering the whole design process. The selected set is as follows:

Stage in the design process	*Method relevant to this stage*
Clarifying objectives	Objectives tree Aim: To clarify design objectives and sub-objectives, and the relationships between them
Establishing functions	Function analysis Aim: To establish the functions required and the system boundary of a new design
Setting requirements	Performance specification Aim: To make an accurate specification of the performance required of a design solution
Generating alternatives	Morphological chart Aim: To generate the complete range of alternative design solutions for a product, and hence to widen the search for potential new solutions
Evaluating alternatives	Weighted objectives Aim: To compare the utility values of alternative design proposals on the basis of performance against differentially weighted objectives

Improving details	Value engineering
	Aim:
	To increase or maintain the value of a product to its purchaser whilst reducing its cost to its producer

As we shall discuss later, in Chapter 10, these six stages of design, and their accompanying design methods, should not be assumed to constitute an invariate design process. However, Figure 21 suggests how they relate to each other and to the symmetrical problem–solution model developed in Chapter 2. For example, clarifying objectives (using the objectives tree method) is appropriate both for understanding the problem–solution relationship and for developing from the overall problem into sub-problems.

In the following six chapters, each of these six methods is presented in a step-by-step procedure, followed by a number of short practical examples and a more complete worked example. The examples show that such methods are often adapted to suit the particular requirements of the task in hand. Although it is important not to follow any method in a slavish and unimaginative fashion, it is also important that an effort is made to follow the principles of the method with some rigour. No beneficial results can be expected from slipshod attempts at 'method'.

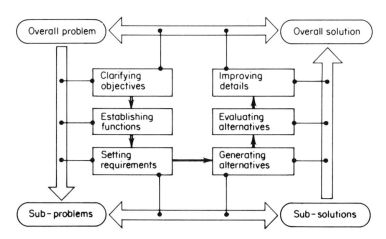

Figure 21. The six stages of the design process positioned within the symmetrical problem–solution model

Chapter 4

CLARIFYING OBJECTIVES

When a client, sponsor or company manager first approaches a designer with a product need, it is unlikely that the 'need' will be expressed very clearly. The client perhaps knows only the type of product that is wanted, and has little idea of the details or of the variants that might be possible. Or the 'need' might be much vaguer still: simply a 'problem' that needs a solution.

The starting point for a design is therefore very often an ill-defined problem or a rather vague requirement. It will be quite rare for a designer to be given a complete and clear statement of design objectives.

Yet the designer must have some objectives to work towards. The outcome of designing is a proposal for some means to achieve a desired end. That 'end' is the set of objectives that the designed object must meet.

An important first step in designing is therefore to try to clarify the design objectives. In fact, it is very helpful at all stages of designing to have a clear idea of the objectives, even though those objectives may change as the design work progresses. The initial and interim objectives may change, expand or contract, or be completely altered as the problem becomes better understood and as solution ideas develop.

It is therefore quite likely that both 'ends' and 'means' will change during the design process. However, as an aid to controlling and managing the design process it is important to have, at all times, a statement of objectives that is as clear as possible. This statement should be in a form that is easily understood and that can be agreed by the client and the designer, or by the various members of the design team. (It is surprising how often members of the same team can have different objectives!)

The *objectives tree* method offers a clear and useful format for such a statement of objectives. It shows the objectives and the general means for achieving them which are under consideration. It shows in a diagrammatic form the ways in which different objectives are related to each other and the hierarchical pattern of objectives and sub-objectives. The procedure for arriving at an objectives tree helps to clarify the objectives and to reach agreement between clients, managers and members of the design team.

The Objectives Tree Method

PROCEDURE

Prepare a list of design objectives

The 'brief' for a design problem is often very aptly called that—it is a very *brief* statement! Such brevity may be because the client is very uncertain about what is wanted or it may be because he or she assumes that the designer perfectly understands what is wanted. Another alternative is that the client wishes to leave the designer with as much freedom as possible. This might sound like a distinct advantage to the designer, but can lead to great frustration when the client decides that the final design proposal is definitely *not* what was wanted! In any case, the designer will almost certainly need to develop the initial brief into a clear statement of design objectives.

The design objectives might also be called client requirements, user needs or product purpose. Whatever they are called, they are the mixture of abstract and concrete aims that the design must try to satisfy or achieve.

Some design objectives will be contained within the design brief; others must be obtained by questioning the client or by discussion in the design team. Typically, initial statements of objectives will be brief and rather vague, such as 'The product must be safe and reliable'. To produce more precise objectives, you will need to expand and to clarify such statements.

One way to begin to make vague statements more specific is, literally, to try to *specify* what it means. Ask 'what is *meant* by that statement'. For example, an objective for a machine tool that it must be 'safe' might be expanded to mean:

1. Low risk of injury to operator
2. Low risk of operator mistakes

3. Low risk of damage to workpiece or tool
4. Automatic cut-out on overload

This kind of list can be generated simply at random as you think about the objective or in discussion within the design team. The client may also have to be asked to be more specific about objectives included in the design brief.

The types of questions that are useful in expanding and clarifying objectives are the simple ones of 'Why?', 'How?' and 'What?'. For instance, ask 'Why do we want to achieve this objective?', 'How can we achieve it?' and 'What implicit objectives underlie the stated ones?' or 'What is the problem really about?'.

Order the list into sets of higher-level and lower-level objectives

As you expand the list of objectives it should become clear that some are at higher levels of importance than others. Sub-objectives for meeting higher-level objectives may also emerge, and some of the statements will be means of achieving certain objectives. This is because some of the questions that you will have been asking about the general objectives imply a 'means-end' relationship— i.e. a lower-level objective is a means to achieving a higher-level one.

An example is the statement 'automatic cut-out on overload' in the list above. This is not really an objective in itself, but a means of achieving an objective—in this case, the objective of 'low risk of damage to workpiece or tool'. In turn, this 'low risk of damage' objective is itself a lower-level objective to that of the overall 'safety' objective.

Your expanded list of objectives will therefore inevitably contain statements at various levels of specificity. In order to clarify the various levels that are emerging, rewrite your general list of objectives into ordered sets; i.e. group the objectives into sets, each concerned with one highest-level objective. For example, one set might be to do with 'safety', another to do with 'reliability', and so on. Within each set, list the sub-objectives in hierarchical order, so that the lower-level ones are clearly separated as means of achieving the higher-level ones. Thus, for instance, your 'safety' list might look like this:

Machine must be safe

Low risk of injury to operator
Low risk of operator mistakes
Low risk of damage to workpiece or tool

Automatic cut-out on overload

The list is now ordered into three hierarchical levels. It can sometimes be difficult to differentiate between levels of objectives, or different people in the design team may disagree about relative levels of importance of some objectives. However, exact precision of relative levels is not important, and you want only a few levels, about which most people can agree. For instance, in the above list, 'low risk of injury' might be considered more important than 'low risk of mistakes', but all three 'low-risk' objectives can conveniently be grouped at about the same level.

The valuable aspect to sorting objectives roughly into levels is that it encourages you to think more clearly about the objectives and about the relationships between means and ends. As you write out your lists in hierarchical levels, you will probably also continue to expand them, as you think of further means to meet sub-objectives to meet objectives, etc.

When you have quite a lot of statements of objectives, it is easier to sort them into ordered sets if each statement is written onto a separate slip of paper or small card. Then you can more easily shuffle them about into groups and levels.

Draw a diagrammatic tree of objectives, showing hierarchical relationships and interconnections

As you write out and shuffle your lists, you will probably realize that some of the sub-objectives relate to, or are means of achieving, more than one higher-level objective. For example, the sub-objective of 'low risk of damage to workpiece or tool' might be not only a means of achieving safety but also a means of achieving reliability.

Therefore a diagram of the hierarchical relationships of these few objectives and sub-objectives might look like Figure 22. This diagram is the beginning of a

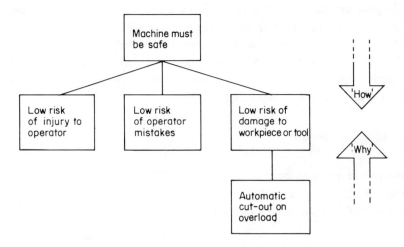

Figure 22. Hierarchical diagram of relationships

'tree' which shows the full pattern of relationships and interconnections. It is not necessarily just a simple 'tree' structure of branches, twigs and leaves, because some of the interconnections form loops or lattices. The 'tree' is also normally drawn 'upside-down'—i.e. usually it has increasingly more 'branches' at lower levels—and so it might be better to think of the sub-objectives as 'roots' rather than 'branches'. It can sometimes be more convenient to draw the 'tree' on its side, i.e. with branches or roots spreading horizontally. In order to help organize the relationships and interconnections between objectives and sub-objectives, draw a complete 'tree' diagram, based on your ordered sets of objectives. Each connecting link that you draw indicates that a lower-level objective is a *means of achieving* the higher-level objective to which it is linked. Therefore working *down* the tree a link indicates *how* a higher-level objective might be achieved; working *up* the tree a link indicates *why* a lower-level objective is included.

Different people might well draw different objectives trees for the same problem or even from the same set of objectives statements. The tree diagram simply represents one perception of the problem structure. The tree diagram helps to sharpen and improve your own perception of the problem or to reach consensus about objectives in a team. It is also only a temporary pattern, which will probably change as the design process proceeds.

As with many other design methods, it is not so much the end product of the method (in this case, the tree diagram) which is itself of most value, but the process of working through the method. The objectives tree method forces you to ask questions about objectives, such as 'What does the client mean by X?'. Such questions help to make the design objectives more explicit and bring them into the open for discussion. Writing the lists and drawing the tree also begins the process of suggesting means of achieving the design objectives, and thus of beginning the process of devising potential design solutions.

Throughout a project, the design objectives should be stated as clearly as the available information permits; the objectives tree facilitates this.

SUMMARY

Objectives tree

Aim

To clarify design objectives and sub-objectives, and the relationships between them.

Procedure
1. Prepare a list of design objectives. These are taken from the design brief, from questions to the client and from discussion in the design team.

2. Order the list into sets of higher-level and lower-level objectives. The expanded list of objectives and sub-objectives is grouped roughly into hierarchical levels.
3. Draw a diagrammatic tree of objectives, showing hierarchical relationships and interconnections. The branches (or roots) in the tree represent relationships that suggest means of achieving objectives.

EXAMPLES

Example 1: City transport system

This is an example of expanding and clarifying design objectives from an initially vague brief. A city planning authority asked a transport design team for proposals for 'a modern system, such as a monorail, which would prevent traffic congestion in the city from getting any worse and preferably remove it altogether'.

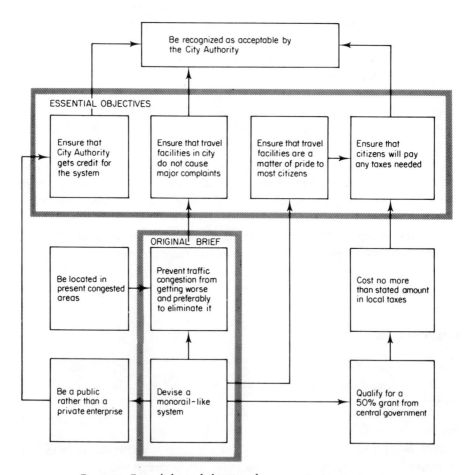

Figure 23. Expanded set of objectives for a new city transport system

The only clear objective in this statement is 'To prevent traffic congestion . . . from getting any worse . . .'. However, what are the implicit objectives behind the desire for 'a modern system, such as a monorail'? Traffic congestion might be held constant or reduced by other means.

By questioning their clients, the design team uncovered objectives such as a desire to generate prestige for the city and to reflect a progressive image for the city authority. There was also a wish simply to reduce complaints from citizens about the existing traffic system. It was also discovered that only certain types of new system would be eligible for a subsidy from central government.

The design team were able to draw up an expanded and hierarchically ordered set of objectives, as shown in Figure 23. In particular, they identified a number of high-level, 'essential objectives' which were not explicitly stated in the original brief. By identifying these objectives, the designers clarifed the project and the limitations that there might be on the range of alternative solutions. (Source: Jones, 1981.)

Example 2: Regional transport system

Another example from transport design is shown in Figure 24, for a larger, regional system. The designers started from the clients' vague definition of 'a convenient, safe, attractive system', and expanded each objective in turn.

For example, 'convenience' was defined in terms of 'low journey times' and 'low "out-of-pocket" costs' for users. The latter objective can be met by appropriate pricing policies; low journey times can be met by a variety of sub-objectives, as shown on the left-hand side of the objectives tree in Figure 24.

Two aspects of 'attractiveness' were defined: user and non-user aspects. The user aspects were subdivided into comfort, visual appeal and internal noise, whereas the non-user aspects were external noise and visual obtrusiveness.

The 'safety' objective was defined to include deaths, injuries and property damage. The sub-objectives to these show how sub-objectives can contribute to more than one higher-level objective. A 'low risk of accidents' can contribute to all three higher-level objectives. If accidents do occur, a 'low risk of injury per accident' can contribute to keeping down both injuries and deaths.

Example 3: Impulse-loading test rig

An example of applying the objectives tree method in engineering design is provided here. The design problem was that of a machine to be used in testing shaft connections subjected to impulse loads.

As before, a typically vague requirement of a 'reliable and simple testing device' can be expanded into a much more detailed set of objectives (Figure 25). 'Reliability' is expanded into 'reliable operation' and 'high safety'. 'Simple' is expanded into 'simple production' and 'good operating characteristics'; the latter is further defined as 'easy maintenance' and 'easy handling'; and so on.

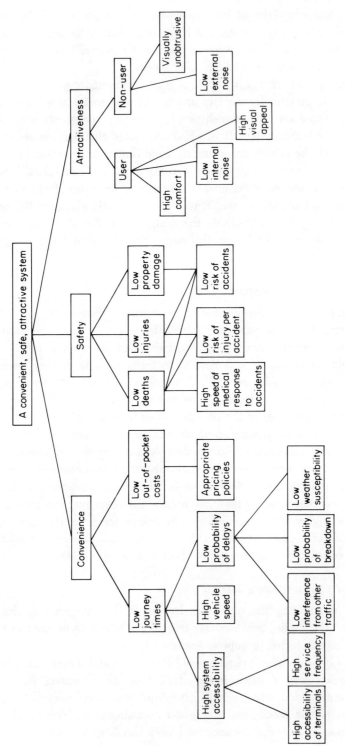

Figure 24. An objectives tree for a 'convenient, safe, attractive' new transport system

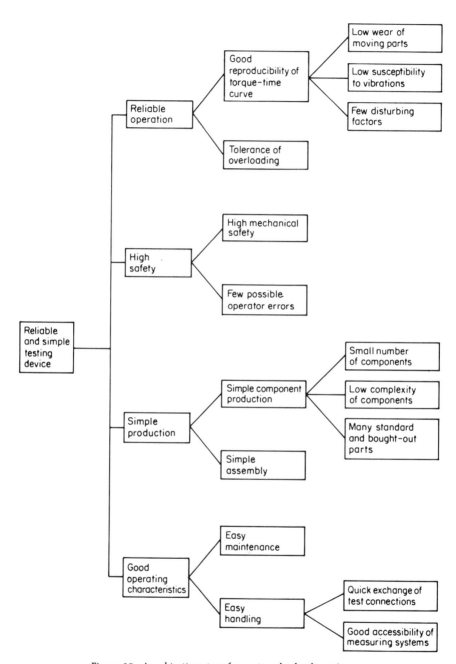

Figure 25. An objectives tree for an impulse-loading test rig

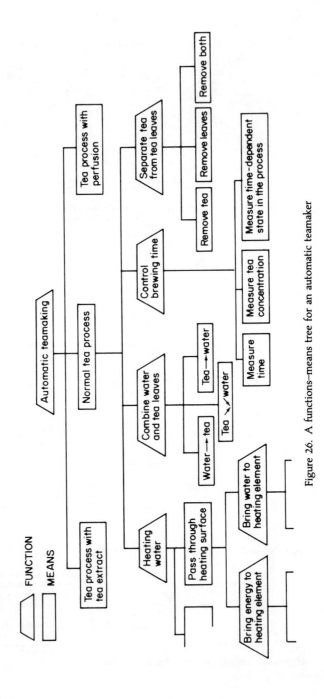

Figure 26. A functions–means tree for an automatic teamaker

In a case such as this, first attempts at expanding the list of objectives would probably produce statements at all levels of generality. For example, asking 'What is meant by "simple"?' would have been likely to produce statements in random order such as 'easy maintenance', 'small number of components', 'simple assembly', etc. Drawing these out in the hierarchical tree structure shows how they relate together. (Source: Pahl and Beitz, 1984.)

Example 4: Automatic teamaker

The objectives tree method can also be used in designing a relatively simple device such as an automatic teamaker. In this example, a distinction is made between 'functions' and 'means'. Each 'function' is an objective, which may be achieved by a number of different 'means' or sub-objectives. Thus the function 'combine water and tea leaves' could be achieved by adding the water to the tea, adding the tea to the water or bringing them both together into one receptacle. (Figure 26.)

This is a variation on the objectives tree as described earlier and demonstrated in the other examples, and might more accurately be called a 'function tree'. However, the same principles apply of breaking down objectives into sub-objectives, or functions into means, and ordering them into a hierarchical tree. This application of the tree structure approach helps to ensure that all the possible means of achieving a function (or objective) are considered by the designer. (Source: Tjalve, 1979.)

Example 5: Railway buffer

Another slightly different use of a 'tree structure' method is this example taken from the design of a railway buffer. As with the teamaker, the tree diagram sets

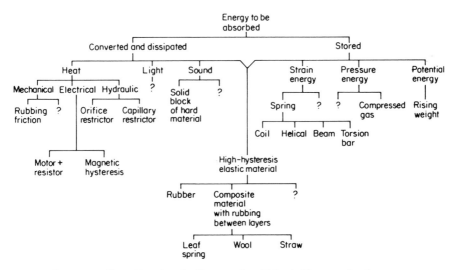

Figure 27. Alternatives tree for the ways in which a buffer may absorb energy

out all the alternative solutions to the problem. The essential function or objective can be subdivided, and sub–subdivided into a large number of alternative means. (Source: Pitts, 1973.)

Worked example: High-pressure pump

This example is based on the design of a pump for high-pressure, high-temperature fluids. The manufacturer who commissioned the design already made a variety of such pumps, but wished to rationalize his range of pumps in order to reduce manufacturing costs. He also wanted to improve the reliability of his pumps and to offer a product that was seen to be convenient to the varied and changing needs of his customers.

On questioning the client about the 'reliability' and 'convenience' objectives, a common aspect emerged: that the pump should be 'robust'—i.e. that it should not easily fail. The initial list of objectives so far might therefore look like this, in hierarchical order:

Reliable
Convenient
Robust
Standardized range

These are all still rather high-level and general objectives, so it is necessary to investigate such statements further. In this case, it was possible to investigate the

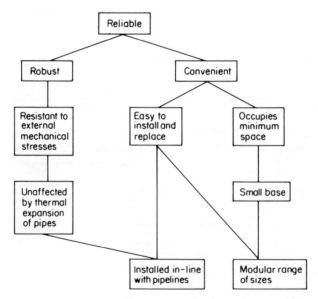

Figure 28. Objectives tree for the pump

problems experienced with the existing pumps. It was found that they were sometimes affected by cracking and leakages due to the stresses caused by the thermal expansion of the pipes to which they were connected. This appeared to be the main problem to which the requests for 'robustness' and 'reliability' were aimed.

Similarly, investigating the 'convenient' objective revealed a further two sub-objectives: firstly, that the pumps should be easy to install and replace and, secondly, that they should occupy the minimum space. It was realized that the standardization of sizes and dimensions in the range could be a means of helping to achieve these objectives, as well as reducing manufacturing costs.

The expanded list of objectives therefore looked like this:

Reliable

Robust	Convenient
Resistant to external mechanical stresses	Easy to install and replace
Unaffected by thermal expansion of pipes	Occupy minimum space
	Standardized range

A key design principle to emerge from considering the means of achieving objectives was that the inlet and outlet ports should always be in-line, to avoid the thermal expansion problems. Such a system, coupled with a small base size and modular dimensioning of alternative components, would also facilitate installation and replacement of the pump. The objectives tree therefore looked like Figure 28.

An actual pump has been designed on similar principles in Denmark (see Figure 29). According to the Danish Design Council, the pump is 'almost a diagram of its problem statement: intake and discharge are aligned, motor, coupling and the stage-built pump are aligned on an axis at right angles to the installation surface, and the pump pressure is increased by adding to the number of stages, i.e. a change in height. The pump is installed directly on the pipeline, occupying a minimum of space.'

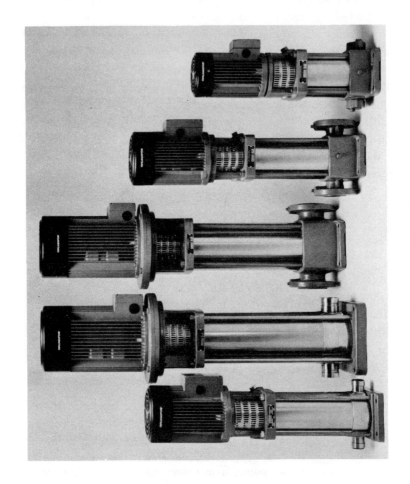

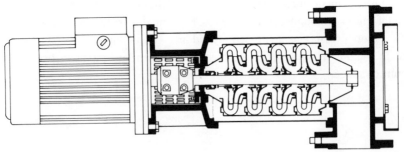

Figure 29. The Danish Grundfos pumps have been designed on similar principles to the objectives developed in this example

Chapter 5

ESTABLISHING FUNCTIONS

We have seen from the objectives tree method that design problems can have many different levels of generality or detail. Obviously, the level at which the problem is defined for or by the designer is crucial. There is a big difference between being asked to 'design a telephone handset' or to 'design a telecommunication system'.

It is always possible to move up or down the levels of generality in a design problem. The classic case is that of the problem 'to design a doorknob'. The designer can move up several levels to that of designing the door or even to designing 'a means of ingress an egress' and find solutions which need no doorknob at all—but this is of no use to a client who manufactures doorknobs! Alternatively, the designer can move down several levels, investigating the ergonomics of handles or the kinematics of latch mechanisms—perhaps again producing non-doorknob solutions which are functional improvements but which are not what the client wanted.

However, there are often occasions when it is appropriate to question the level at which a design problem is posed. A client may be focusing too narrowly on a certain level of problem definition, when a resolution at another level might be better, and reconsidering the level of problem definition is often a stimulus to the designer to propose more radical or innovative types of solutions.

It is useful, therefore, to have a means of considering the problem level at which a designer or design team is to work. It is also very useful if this can be done in a way that considers, not the potential type of solution, but the essential functions that a solution type will be required to satisfy. This leaves the designer free to develop alternative solution proposals that satisfy the functional requirements.

The *function analysis* method offers such a means of considering essential functions and the level at which the problem is to be addressed. The essential functions are those that the device, product or system to be designed must satisfy, no matter what physical components might be used. The problem level is decided by establishing a 'boundary' around a coherent sub-set of functions.

The Function Analysis Method

PROCEDURE

Express the overall function for the design in terms of the conversion of inputs and outputs

The starting point for this method is to concentrate on *what* has to be achieved by a new design, and not on *how* it is to be achieved. The simplest and most basic way of expressing this is to represent the product or device to be designed as simply a 'black box' which converts certain 'inputs' into desired 'outputs'. The 'black box' contains all the functions that are necessary for converting the inputs into the outputs (Figure 30).

It is preferable to try to make this overall function as broad as possible at first—it can be narrowed down later if necessary. It would be wrong to start with an unnecessarily limited overall function which restricts the range of possible solutions. The designer can make a distinct contribution to this stage of the design process by asking the clients or users for definitions of the fundamental purpose of the product or device, and asking about the required inputs and outputs—from where do the inputs come, what are the outputs for, what is the next stage of conversion, etc.?

This kind of questioning is known as 'widening the system boundary'. The 'system boundary' is the conceptual boundary that is used to define the function of the product or device. Often, this boundary is defined too narrowly, with the

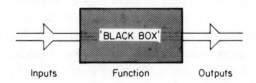

Inputs   Function   Outputs

Figure 30. The 'black box' systems model

result that only minor design changes can be made, rather than a radical rethinking.

It is important to try to ensure that all the relevant inputs and outputs are listed. They can all usually be classified as flows of either materials, energy or information, and these same classifications can be used to check if any input or output type has been omitted.

Break down the overall function into a set of essential sub-functions

Usually, the conversion of the set of inputs into the set of outputs is a complex task inside the 'black box', which has to be broken down into sub-tasks or sub-functions. There is no really objective, systematic way of doing this; the analysis into sub-functions may depend on factors such as the kinds of components available for specific tasks, the necessary or preferred allocations of functions to machines or to human operators, the designer's experience, and so on.

In specifying sub-functions it is helpful to ensure that they are all expressed in the same way. Each one should be a statement of a verb plus a noun—e.g 'amplify signal', 'count items', 'separate waste', 'reduce volume'.

Each sub-function has its own input(s) and output(s), and compatibility between these should be checked. There may be 'auxiliary sub-functions' that have to be added but which do not contribute directly to the overall function, such as 'remove waste'.

Draw a block diagram showing the interactions between sub-functions

A block diagram consists of all the sub-functions separately identified by enclosing them in boxes and linked together by their inputs and outputs so as to satisfy the overall function of the product or device that is being designed. In other words, the original 'black box' of the overall function is redrawn as a 'transparent box' in which the necessary sub-functions and their links can be seen (Figure 31).

In drawing this diagram you are deciding how the internal inputs and outputs of the sub-functions are linked together so as to make a feasible, working system. You may find that you have to juggle inputs and outputs, and perhaps redefine some sub-functions so that everything is connected together. It is useful to use

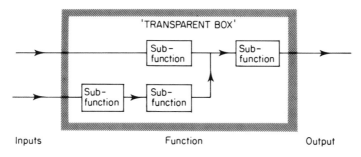

Figure 31. A 'transparent box' model

different conventions—i.e. different types of lines—to show the different types of inputs and outputs—i.e. flows of materials, energy or information.

Draw the system boundary

In drawing the block diagram you will also need to make decisions about the precise extent and location of the system boundary. For example, there can be no 'loose' inputs or outputs in the diagram except those that come from or go outside the system boundary.

It may be that the boundary now has to be narrowed again, after its earlier broadening during consideration of inputs, outputs and overall function. The boundary has to be drawn around a sub-set of the functions that have been identified, in order to define a feasible product. It is also probable that this drawing of the system boundary is not something in which the designer has complete freedom—as likely as not, it will be a matter of management policy or client requirements. Usually, many different system boundaries can be drawn, defining different products or solution types.

Search for appropriate components for performing the sub-functions and their interactions

If the sub-functions have been defined adequately and at an appropriate level, then it should be possible to identify a suitable component for each sub-function. This identification of components will depend on the nature of the product or device, or more general system, that is being designed. For instance, a 'component' might be defined as a person who performs a certain task, a mechanical component, or an electronic device. One of the interesting design possibilities opened up by electronic devices such as microprocessors is that these can often now be substituted for components that were previously mechanical devices or perhaps could only be done by human operators. The function analysis method is a useful aid in these circumstances because it focuses on functions, and leaves the physical means of achieving those functions to this later stage of the design process.

SUMMARY

Function analysis

Aim
To establish the functions required, and the system boundary, of a new design.

Procedure
1. Express the overall function for the design in terms of the conversion of inputs into outputs. The overall, 'black box' function should be broad—widening the system boundary.

2. Break down the overall function into a set of essential sub-functions. The sub-functions comprise all the tasks that have to be performed inside the 'black box'.

3. Draw a block diagram showing the interactions between sub-functions. The 'black box' is made 'transparent', so that the sub-functions and their interconnections are clarified.

4. Draw the system boundary. The system boundary defines the functional limits for the product or device to be designed.

5. Search for appropriate components for performing the sub-functions and their interactions. Many alternative components may be capable of performing the identified functions.

EXAMPLES

Example 1: Feed delivery system

The function analysis method is particularly relevant in the design of flow-process systems, such as that shown diagrammatically in Figure 32. This represents a factory where animal feedstuffs are bagged.

In this example, the company wanted to try to reduce the relatively high costs of handling and storing the feedstuffs. A designer might tackle this task by searching for very direct ways in which each part of the existing process might be made cost-effective. However, a broader formulation of the problem—the overall function—was represented in the following stages:

1. Transfer of feed from mixing bin to bags stored in warehouse
2. Transfer of feed from mixing bin to bags loaded on truck
3. Transfer of feed from mixing bin to consumer's storage bins
4. Transfer of feed ingredients from source to consumer's storage bins

This broadening of the problem formulation is shown diagrammatically in Figure 33.

Each different formulation suggests different kinds of solutions, with the broadest formulation perhaps leading to the complete elimination of the handling, storing and loading sub-functions. (Source: Krick, 1976.)

Example 2: Packing carpet squares

This example shows another flow process—the packing of loose carpet squares into lots. The designers first broke down the overall function into a series of principal sub-functions (Figure 34). Some auxiliary functions then became clear. For example, the input from the separate stamping machine includes offcuts which have to be removed; reject squares must also be removed; materials must be

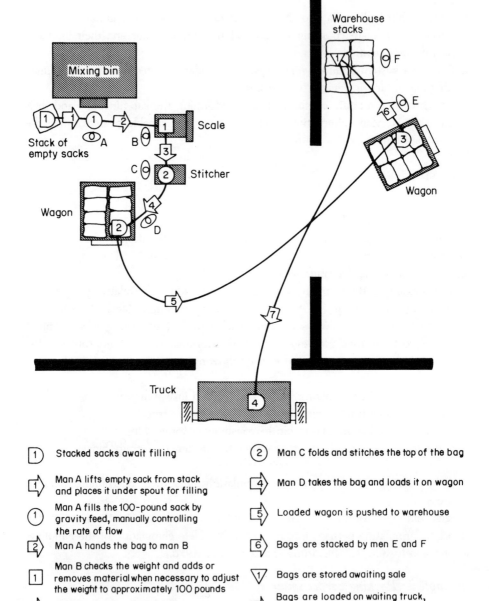

Figure 32. The existing method of filling, storing and despatching bags of animal feed

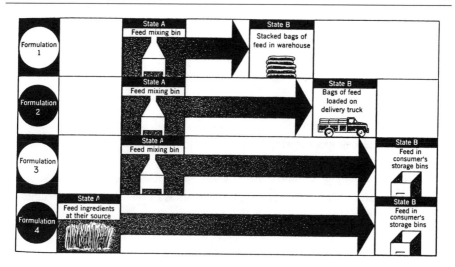

Figure 33. Alternative formulations of the feed distribution problem

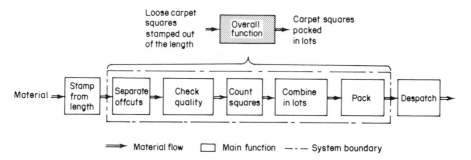

Figure 34. Analysis of principal functions for the packing of carpet squares

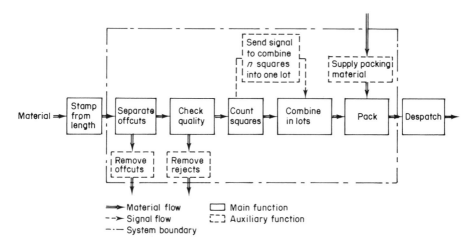

Figure 35. Expanded function analysis for the packing of carpet squares

brought in for packaging. The sub-function 'count squares' could also be used to give the signal for packaging lots of a specified number (see Figure 35). (Source: Pahl and Beitz, 1984.)

Example 3: Fuel gauge

Function analysis can also be applied in the design of much smaller products or devices. Figure 36 shows the step-by-step development of a function analysis for a fuel gauge. Notice how auxiliary functions are introduced in order to cope with a gradually broadening problem formulation to provide for fuel containers of different sizes and shapes, etc. The figure also shows how the system boundary can be drawn in different places, in this case depending on whether the output signal is to be to already-existing instruments or whether such an instrument is to be included as an integral part of the design. (Source: Pahl and Beitz, 1984.)

Example 4: Flexible coupling

The relationship between a function analysis and a physical product is shown in the example of Figure 37. The product is a coupling, which consists of two 'sub-systems', a flexible coupling and a clutch. Although the function analysis might stay the same, clearly the choice and design of the components that make up the product could vary considerably. (Source: Pahl and Beitz, 1984.)

Worked example: Washing machine

A relatively simple example of the use of the function analysis method is based on the domestic washing machine. The overall function of such a machine is to convert an input of soiled clothes into an output of clean clothes, as shown in Figure 38.

Inside the 'black box' there must be a process that separates the dirt from the clothes, and so the dirt itself must also be a separate output. We know that the conventional process involves water as a means of achieving this separation, and that a further stage must therefore be the conversion of clean (wet) clothes to clean (dry) clothes. Even further stages involve pressing and sorting clothes. The inputs and outputs might therefore be listed like this:

Inputs	*Outputs*
Soiled clothes	(Stage 1) Clean clothes
	Dirt
	(Stage 2) Dry clothes
	Water
	(Stage 3) Pressed clothes

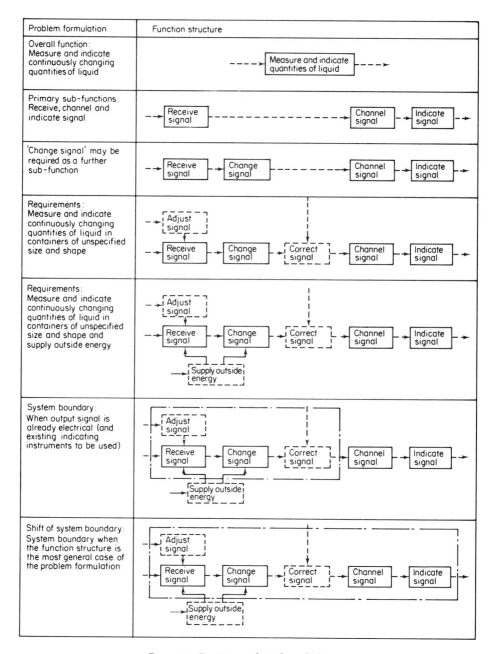

Figure 36. Function analysis for a fuel gauge

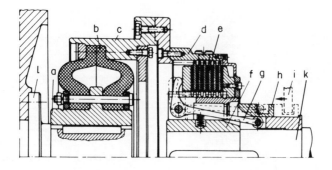

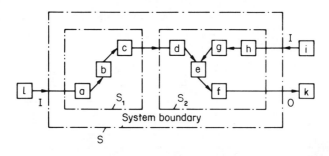

a...h   System elements
i...l   Connecting elements
S   Overall system
S_1 Sub-system : flexible coupling
S_2 Sub-system : clutch
I   Inputs
O   Outputs

Figure 37. Function analysis of a coupling

The essential sub-functions, together with the conventional means of achieving them, for converting soiled clothes into clean, pressed clothes would therefore be as follows:

Essential sub-functions	*Means of achieving sub-functions*
Loosen dirt	Add water and detergent
Separate dirt from clothes	Agitate
Remove dirt	Rinse
Remove water	Spin
Dry clothes	Blow with hot air
Smooth clothes	Press

A block diagram, with main and subsidiary inputs and outputs might look like Figure 39.

The development of washing machines has involved progressively widening

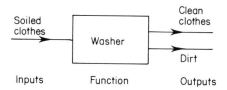

Figure 38. Black box model of a washing machine

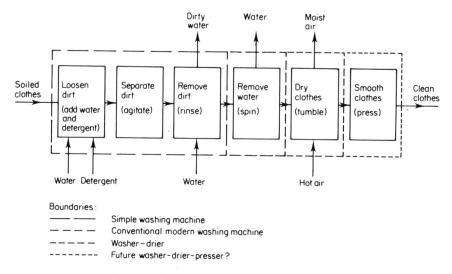

Figure 39. Function analysis of the washing machine

the system boundary, as shown in the figure. Early washing machines simply separated the dirt from the clothes, but did nothing about removing the excess water from the clothes; this was left as a task for the human operator, using either hand or mechanical wringing of the clothes. The inclusion of a spin-drying function removed the excess water, but still left a drying process. This is now incorporated in washer–driers. Perhaps the smoothing of clothes will somehow be incorporated in future machines? (However, this need has been reduced by the use of artificial fabrics in clothes.)

Chapter 6

SETTING REQUIREMENTS

Design problems are always set within certain limits. One of the most important limits, for example, is that of cost: what the client is prepared to spend on a new machine or what customers may be expected to pay as the purchase price of a product. Other common limits may be the acceptable size or weight of a machine; some limits will be performance requirements, such as an engine's power rating; still others might be set by statutory legal or safety requirements.

This set of requirements comprises the performance specification of the product or machine. Statements of design objectives or functions (such as those derived from objectives tree or function analysis methods) are sometimes regarded as being performance specifications, but this is not really correct. Objectives and functions are statements of what a design must achieve or do, but they are not normally set in terms of precise limits, which is what a performance specification does.

In setting limits to what has to be achieved by a design a performance specification thereby limits the range of acceptable solutions. Because it therefore sets the designer's target range, it should not be defined too narrowly. If it is, then a lot of otherwise acceptable solutions might be eliminated unnecessarily. On the other hand, a specification that is too broad or vague can leave the designer with little idea of the appropriate direction in which to aim. Specification limits that are set too wide can also lead to inappropriate solutions which then have to be changed or modified when it is found that they actually fall outside acceptable limits.

There are thus good reasons for putting some effort into an accurate perform-ance specification early in the design process. Initially, it sets up some boundaries to the 'solution space' within which the designer must search. Later on in the design process, the performance specification can be used in evaluating proposed solutions, to check that they do fall within the acceptable boundaries.

The *performance specification* method is intended to help in defining the design problem, leaving the appropriate amount of freedom so that the designer has room to manoeuvre over the ways and means of achieving a satisfactory design solution. A specification defines the required *performance*, and not the required *product*. The method therefore emphasizes the performance that a design solution has to achieve and not any particular physical components which may be means of achieving that performance.

The Performance Specification Method

PROCEDURE

Consider the different levels of generality of solution that might be applicable

It is important that a specification is addressed to an appropriate level of generality for the solution type that is to be considered. A specification at too high a level of generality may allow inappropriate solutions to be suggested, whereas too low a level (a specification that is too specific) can remove almost all of the designer's freedom to generate a range of acceptable solutions.

So the first step is to consider the different levels of generality. A simple classification of types of level, from the most general down to the least, for a product might be:

Product *alternatives*
Product *types*
Product *features*

As an example to illustrate these levels, suppose that the product in question is a domestic heating appliance. At the highest level of generality the designer would be free to propose alternative ways of heating a house, such as moveable appliances, fixed appliances, central heating with radiators, ducted warm air, etc. There might even be freedom to move away from the concept of an 'appliance' to alternative forms of heating such as conservatories that trap solar heat; or to ways of retaining heat, such as insulation. At the intermediate level, the designer would have a much more limited freedom, and might only be concerned with different types of appliance, say different heater types such as radiators or convectors, or different fuel types. At the lowest level, the designer would be constrained to considering different features within a particular type of appliance, such as its heating element, switches, body casing, supports, etc.

Determine the level of generality at which to operate

Considering the different levels of generality might lead either to a broadening or a narrowing of initial product concepts or of the design brief. The second step of the method is therefore to make a decision on the appropriate level.

Normally, the client, company management or customer decides the level at which the designer will operate. For instance, in the case of domestic heating appliances, the highest level of generality ('alternatives') would only be considered if an appliance manufacturer was proposing to diversify or broaden its activities into other aspects of domestic heating. The intermediate level ('types') would normally be considered when a new product was to be designed, to add to the existing range of appliances or to replace obsolete ones. The lowest level ('features') would be considered when making modifications to existing products.

The higher the level of generality that may be considered, then the more freedom the designer has in terms of the range of acceptable solutions. Of course, the higher levels also subsume the lower levels of specification—i.e. the specification of features is part of the specification of types which is part of the specification of alternatives.

Identify the required performance attributes

Once the level at which designing is to proceed has been decided, work can begin on the performance specification proper. Any product or machine will have a set of *attributes*, and it is these that are specified in the performance specification. Attributes include such things as weight, size, colour, speed, cost, life expectancy and key features such as safety, comfort, portability and operating sequence.

Performance attributes are usually similar to, or derived from, the design objectives and functions. Therefore, if you have already prepared an objectives tree or a function analysis, these are likely to be the source of your initial list of performance attributes.

A most important aspect to bear in mind when listing performance attributes is that they should be stated in a way that is independent of any particular solution. Statements of attributes made by clients or customers are often couched in terms of solutions, because they value some performance aspect that is embodied in the solution but have not separated the attribute from a particular embodiment. Such solution-based, rather than performance-based, statements are usually unnecessarily restrictive of solution concepts.

For example, a client might suggest that the material for a particular surface area should be ceramic, because that is a satisfactory feature of an existing solution, but the essential performance requirement might be that the surface should be non-porous, or easy to clean, or have a smooth hard texture, or simply have a shiny appearance. Acceptable alternatives might be plastics, metal or marble.

There may be a whole complex of reasons underlying a client or customer specification of a particular solution feature. It could be the whole set of attributes of a ceramic surface, as just listed, plus the mass which is provided by a ceramic component, plus the colour range, plus some perceived status or other value which is not immediately obvious. A comprehensive and reliable list of performance attributes can therefore take some considerable effort to compile, and may well require careful research into client, customer and perhaps manufacturer requirements.

The final list of performance attributes contains all the conditions that a design proposal should satisfy. However, it may become necessary to distinguish within this list between those attributes or requirements that are 'demands' and those that are 'wishes'. 'Demands' are requirements that *must* be met, whereas 'wishes' are those that the client, customer or designer would like to meet if possible . For example, the requirement of a non-porous surface might be a functional 'demand', but availability in a range of colours might be a 'wish' dependent on the material actually chosen.

State succinct and precise performance requirements for each attribute

Once a reliable list of attributes has been compiled, a performance specification is written for each one. A specification says what a product must *do*, not what it must *be*. Again, this may well require some careful research—it is not adequate simply to guess at performance requirements, nor just to take them from an existing solution type.

Wherever possible, a performance specification should be expressed in quantified terms. Thus, for example, a maximum weight should be specified, rather than a vague statement such as 'lightweight'. A safety specification—say, for escape from a vehicle—should state the maximum time allowable for escape in an emergency, rather than using terms like 'rapidly' or 'readily'.

Also wherever possible and appropriate, a specification should set a range of limits within which acceptable performance lies. So a specification should not say 'Seat height: 425 mm' if a range between 400 and 450 mm is acceptable. On the other hand, spurious 'precision' is also to be avoided: do not specify 'a container of volume 21.2 l' if you mean to refer to a wastepaper bin of approximately 300 mm diameter and 300 mm high'.

SUMMARY

Performance specification

Aim

To make an accurate specification of the performance required of a design solution.

Procedure

1. Consider the different levels of generality of solution that might be applicable. There might be a choice between
   (a) product alternatives,
   (b) product types,
   (c) product features.
2. Determine the level of generality at which to operate. This decision is usually made by the client. The higher the level of generality, the more freedom the designer has.
3. Identify the required performance attributes. Attributes should be stated in terms that are independent of any particular solution.
4. State succinct and precise performance requirements for each attribute. Wherever possible, specifications should be in quantified terms and identify ranges between limits.

EXAMPLES

Example 1: Fuel gauge

This example continues the fuel gauge problem started in the previous chapter. The problem was formulated by the client at the lowest level of generality: the design of a particular type of fuel gauge for use in motor vehicles. The initial general formulation of the problem statement was:

> A gauge to measure continuously changing quantities of liquid in containers of unspecified size and shape, and to indicate the measurement at various distances from the containers.

The following list of attributes was then developed:

> Suitable for containers (fuel tanks) of
>   various volumes
>   various shapes
>   various heights
>   various materials

Specification

for Fuel gauge *Page* 1

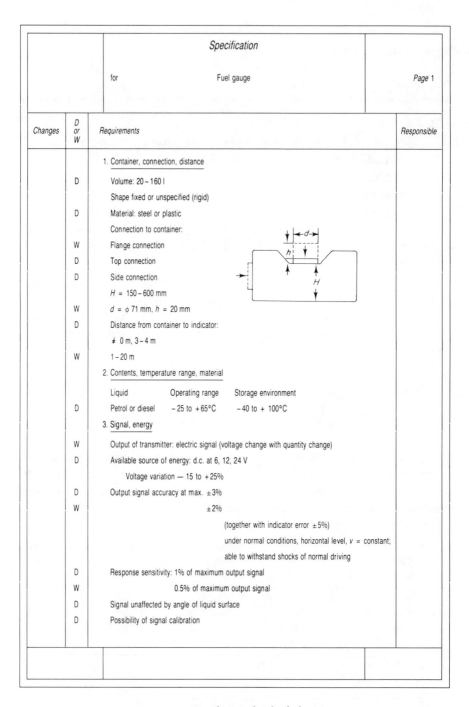

Changes	*D or W*	*Requirements*	*Responsible*
		1. Container, connection, distance	
	D	Volume: 20 – 160 l	
		Shape fixed or unspecified (rigid)	
	D	Material: steel or plastic	
		Connection to container:	
	W	Flange connection	
	D	Top connection	
	D	Side connection	
		H = 150 – 600 mm	
	W	d = ϕ 71 mm, h = 20 mm	
	D	Distance from container to indicator:	
		≠ 0 m, 3 – 4 m	
	W	1 – 20 m	
		2. Contents, temperature range, material	
		Liquid — Operating range — Storage environment	
	D	Petrol or diesel — – 25 to + 65°C — – 40 to + 100°C	
		3. Signal, energy	
	W	Output of transmitter: electric signal (voltage change with quantity change)	
	D	Available source of energy: d.c. at 6, 12, 24 V	
		Voltage variation — 15 to + 25%	
	D	Output signal accuracy at max. ± 3%	
	W	± 2%	
		(together with indicator error ± 5%) under normal conditions, horizontal level, v = constant; able to withstand shocks of normal driving	
	D	Response sensitivity: 1% of maximum output signal	
	W	0.5% of maximum output signal	
	D	Signal unaffected by angle of liquid surface	
	D	Possibility of signal calibration	

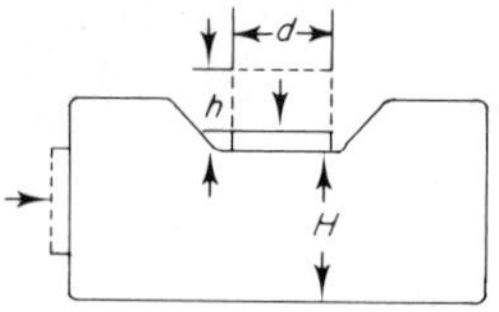

Figure 40a. Specification for the fuel gauge

Changes	D or W	Requirements	Responsible
		Specification	
		for Fuel gauge	*Page 2*
	W	Possibility of signal calibration with full container	
	D	Minimum measurable content: 3% of maximum value	
	W	Reserve tank contents by special signal	
		4. Operating conditions	
	D	Forward acceleration ±10 m/s^2	
	D	Sideways acceleration ±10 m/s^2	
	D	Upward acceleration (vibration) up to 30 m/s^2	
	W	Shocks in forward direction without damage up to 30 m/s^2	
	D	Forward tilt up to ±30°	
	D	Sideways tilt max. 45°	
	D	Tank not pressurized (ventilated)	
		5. Test requirements	
	D	Salt spray tests for inside and outside components according to client's requirements	
	D	Pressure test for container 30 kN/m^2	
		6. Life expectancy, durability of container	
	D	Life expectancy 5 years in respect of corrosion due to contents and condensation	
	D	Must conform with heavy vehicle requirements	
		7. Production	
	W	Simply modified to suit different container sizes	
		8. Operation, maintenance	
	W	Installation by non-specialist	
	D	Must be replaceable and maintenance-free	
		9. Quantity	
		10 000/day of the adjustable type, 5000/day of the most popular type	
		10. Costs	
		Manufacturing costs ⩽ DM 3.00 each	

Figure 40b.

Connection to top or side of container
Operates at various distances from container
Measures petrol or diesel liquid
Accurate signal
Reliable operation

The design team went on to develop a full performance specification, as shown in Figure 40. They also distinguished between 'demands' (D) and 'wishes' (W). (Source: Pahl and Beitz, 1984.)

Example 2: One-handed mixing tap

This example is a specification for a domestic water mixing tap that can be operated with one hand (Figure 41). 'Demands' and 'wishes' in the specification are indicated as in the previous example. (Source: Pahl and Beitz, 1984.)

Example 3: Electric toothbrush

This example shows the development of a performance specification for a consumer product: an electric toothbrush. The problem is set at the intermediate level of generality, i.e. a new type of toothbrush, but it has novel features which require precise performance specifications. The designers listed the new product's attributes mainly in terms of a set of 'user needs':

User needs at the concept phase

Physiological needs	Clean teeth better than a handbrush, massage gums, reduce decay, hygienic family sharing, electrical and mechanical safety, etc.
Social needs	Sweet breath and white teeth (symbolic needs for social acceptance); handle colours to match bathroom, etc.
Psychological needs	Autonomy in deciding when and how one's teeth are to be cared for, self-esteem from care of teeth, praise for effort, pleasure from giving or receiving a gift, etc.
Technical needs	Diameter, length, brush size, amplitude, frequency, weight, running time, reliability, useful life, etc.
Time needs	Needed for Christmas market
Resources exchanged	$1 per person is the lowest cost alternative, but electric razors sell for twenty times the price of a manual razor, so probably $20 will be paid for an electric toothbrush

		Specification		
		for	One-handed mixing tap	*Page* 1

Changes	D or W	Requirements	Responsible
	D	1 Throughput (mixed flow) max. 10 l/min at 2 bar	
	D	2 Max. pressure 10 bar (test pressure 15 bar as per DIN 2401)	
	D	3 Temp. of water: standard 60°C, 100°C (short-time)	
	D	4 Temperature setting independent of throughput and pressure	
	W	5 Permissible temp. fluctuation ±5°C at a pressure diff. of ±5 bar between hot and cold supply	
	D	6 Connection: 2 × Cu pipes, 10 × 1 mm, l = 400 mm	
	D	7 Single-hole attachment $\phi\ 35^{+2}_{-1}$ mm, basin thickness 0 – 18 mm (Observe basin dimensions DIN EN 31, DIN EN 32, DIN 1368)	
	D	8 Outflow above upper edge of basin: 50 mm	
	D	9 To fit household basin	
	W	10 Convertible into wall fitting	
	D	11 Light operation (children)	
	D	12 No external energy	
	D	13 Hard water supply (drinking water)	
	D	14 Clear identification of temperature setting	
	D	15 Trade mark prominently displayed	
	D	16 No connection of the two supplies when valve shut	
	W	17 No connection when water drawn off	
	D	18 Handle not to heat above 35°C	
	W	19 No burns from touching the fittings	
	W	20 Provide scalding protection if extra costs small	
	D	21 Obvious operation, simple and convenient handling	
	D	22 Smooth, easily cleaned contours, no sharp edges	
	D	23 Noiseless operation (≤20 dB as per DIN 52218)	
	W	24 Service life 10 years at about 300 000 operations	
	D	25 Easy maintenance and simple repairs. Use standard spare parts	
	D	26 Max. manuf. costs DM 30 (3000 units per month)	
	D	27 Schedules from inception of development	

	Conceptual design	Embodiment design	Detail Design	Prototype	
after	2	4	6	9	months

Figure 41. Specification for a one-handed mixing tap

The performance specification was then drawn up as a set of design objectives with corresponding criteria, as shown in Figure 42. (Source: Love, 1980.)

Example 4: City car

In this example, a design team undertook a study of the 'city car' concept—i.e. a small runabout for use in cities or other limited-journey purposes. Many different solutions have been developed to this problem, with varying degrees of success.

The design study included an analysis of the features of the many previous examples of city car designs, as well as market research, town planning and engineering criteria, etc. As a result, a specification of the general features required

Objectives	Criteria
1. To be attractive, suitable for sale primarily in the gift market and secondly as a personal purchase.	1a. Attractiveness of overall design and packaging to be judged better than brands X and Y by more than 75% of a representative consumer panel.
	1b. Decorator colours to be the same as our regular products.
	1c. Package can be displayed on counter area of 75 × 100 mm.
2. The technical functions are to be at least as good as past 'family' models of brand X.	2a. Technical functions to be judged at least as good as the past 'family' model of brand X by dental consultant, Dr J.P.
	2b. Amplitude to be between 2 and 3 mm.
	2c. Frequency to be 15 ± 5 cycles/s.
	2d. Battery life to be minimum of 50 min. when tested according to standard XYZ.
	2e, etc., for other technical aspects such as weight, impact strength, frequency of repair, dimensions . . . .
3. To be saleable in the United States and Canada.	3. Must meet UL and CSA standards for safety (a crucial criterion).
4. The timing objective is that the product be ready for sale to the Christmas trade in the nearest feasible season.	4. The time milestones, backing up from October production are to be: • mock-up approval—2 months • tooling release—6 months • production prototype—10 months • pilot run—10 months • production run—13 months (October).
5. The selling price is to be not more than 10% of the present utility models.	5. The selling price is to be between $12.50 and $17.50, depending on the features offered, for a production run of 100 000 units.

Figure 42. Performance specification for an electric toothbrush

or desired in a city car was drawn up and is shown in Figure 43. The characteristics are classified as either 'required' or 'desired'—similar to 'demands' and 'wishes' in other examples. (Source: Pighini *et al.*, 1983.)

Worked example: Portable computer

Personal computers have become a fundamental aspect of many activities. Some people need to have not only one at the office and one at home but also one that can be used at other locations and that therefore travels with them. This example is based on the design of such a portable computer.

Clearly, the level of generality for this problem has been set by the client's request for a new design of portable computer: it is a particular product *type*, with the designers' freedom therefore constrained to product *features*.

There are many specialized attributes which would have to be researched and specified, such as the type of microprocessors to be incorporated, type of screen display, keyboard, etc. We shall concentrate here only on the key attribute of 'portable'. What exactly does this mean? We need to know what features of 'portability' might be important to potential purchasers of the computer.

We therefore interview a range of computer users about their needs. From this it emerges that there are two distinct aspects to 'portability'. The first is, quite simply, that the machine can be carried comfortably and easily—in contrast to some, which are known as 'luggable' rather than 'portable'! The second aspect is that the purpose of a portable machine is that it can be used in a wide variety of different locations.

Further research with users is necessary to develop performance specifications for both of these attributes. For example, to specify the 'carryable' performance attribute, it is not adequate simply to suggest a carrying handle. Nor is it adequate just to weigh a rival product and specify that as maximum weight. We need to know the range of users for the computer and the typical distances or lengths of time that it might be carried. Experiments with a few representative, least-strong users and maximum expected journeys could then establish an appropriate weight limit.

We also need to investigate further the variety of locations in which it is desired to use the computer. Does this include someone's lap on a train or plane? In that case the machine must be small but stable. Does it include word-processing use during meetings, conferences or lectures? In that case its operation should be silent or very quiet. Does it include out-of-doors use? In that case there might be weatherproofing requirements, or the user might be wearing gloves.

Obviously, in many locations there is no available power source, and so a portable computer must have its own batteries. It may be that salespeople and others often use the computer in their car; therefore the car's battery could be used through the cigar-lighter socket. One other aspect of performance that emerges is that users want to be able to plug the portable computer into

	Characteristics	
	Required	Desired
1 General characteristics		
Car for city or for delimited area (airport, port, railway station, industry, etc.)	X	
Number of seats: 2(4)	X	
Number of wheels: 4	X	
Utilization of space: maximum in relation to the external dimensions		X
Vehicle must be economical	X	
Range > 100 km		X
2 Working conditions		
The same as those relative to the use of every automobile in city or for delimited area	X	
3 Dimensions		
Length 2.5 m	X	
Width 1.5 m		X
Height 1.6 m		X
4 Weight		
Maximum net weight 400 kg		X
Available loading capacity 200 kg (300 kg)	X	
Gross weight 600 kg (700 kg)		X
5 Capacity for luggage		
Minimum volume 150 dm^3	X	
With back row seats dropped down 350 dm^3		X
Possibility for loading outside (on the roof)		X
6 Velocity		
Average velocity 50 km/h	X	
Maximum velocity 70 km/h	X	
7 Type of motor		
Thermal (internal combusion, etc.)	X	
Electrical		
Thermoelectrical		
Thermohydraulic		
Air compressed		
8 Characteristics of motor		
Power: 2 kW		X
Fuel consumption 1 litre per 25 km		X
(thermic motor)		
9 Maintenance		
Minimum and simple		X
Possibility of finding spare parts: easy		X
10 Utilization, use		
Utilization: simple		X
Use: frequent	X	
Reliability: high	X	
11 Durability		
At least 50 000 km or 5 years	X	
12 Safety		
As high as possible (active and passive)	X	
13 Pollution		
Not above the legal minimum	X	
Zero		X
14 Form and aesthetics		
Pleasant and therefore able to be commercialized	X	
Convertible		X
15 Production		
In small series (500 vehicles/year)		X
16 Price		
$2000 to $2500 (2/3 Fiat 126)	X	

Figure 43. Specification for a small city car

Figure 44. The Amstrad PPC portable computer was designed to a 'portability' specification similar to that developed in this example

conventional devices such as VDUs and printers at different locations in order to provide a display visible by a group of people or to leave a hard-copy printout. This means that hardware compatibility with such devices is important.

A performance specification for the 'portability' attribute is therefore developed as follows:

Performance specification   Portable computer
Portability
    Can be carried in one hand
    Weight not more than 6 kg
    Optional carrying case, with pockets for disks
    When closed, weatherproof against rain showers
    Maximum base dimensions in use: 500×500 mm
    Near-silent keyboard
    Power sources:
        mains
        own battery
        car cigar-lighter socket
    Compatible with standard VDUs and printers

Chapter 7

GENERATING ALTERNATIVES

The generation of solutions is, of course, the essential, central aspect of designing. Whether one sees it as a mysterious act of creativity or as a logical process of problem-solving, the whole purpose of design is to make a proposal for something new—something that does not yet exist.

The focus of much writing and teaching in design is therefore on novel products or machines, which often appear to have arisen spontaneously from the designer's mind. However, this overlooks the fact that most designing is actually a variation from or modification to an already-existing product or machine. Clients and customers usually want improvements rather than novelties.

Making variations on established themes is therefore an important feature of design activity. It is also the way in which much creative thinking actually develops. In particular, creativity can often be seen as the reordering or recombination of existing elements.

This creative reordering is possible because even a relatively small number of basic elements or components can usually be combined in a large number of different ways. A simple example of arranging adjacent squares into patterns demonstrates this:

No. of squares	No. of distinct shape arrangements
2	1
3	2
4	5
5	12
6	35
7	108
8	369
.	.
.	.
.	.
16	13 079 255

The number of different arrangements—i.e. patterns or designs—soon becomes a 'combinatorial explosion' of possibilities.

The *morphological chart* method exploits this phenomenon and encourage the designer to identify novel combinations of elements or components. The chart sets out the complete range of elements, components or sub-solutions that can be combined together to make a solution. The number of possible combinations is usually very high, and includes not only existing, conventional solutions but also a wide range of variations and completely novel solutions.

The main aim of this method is to widen the search for possible new solutions. 'Morphology' means the study of shape or form, so a 'morphological analysis' is a systematic attempt to analyse the form that a product or machine might take and a 'morphological chart' is a summary of this analysis. Different combinations of sub-solutions can be selected from the chart, perhaps leading to new solutions that have not previously been identified.

The Morphological Chart Method

PROCEDURE

List the features or functions that are essential to the product

The purpose of this list is to try to establish those essential aspects that must be incorporated in the product or that it must be capable of doing. These are therefore usually expressed in rather abstract terms of product requirements or functions. In the morphological chart method they are sometimes called the design 'parameters'. As with many other design methods, instead of thinking in terms of the physical components that a typical product might have, you have to think of the functions that those components serve.

The items in the list should all be at the same level of generality, and they should be as independent of each other as possible. They must also comprehensively cover the necessary functions of the product or machine to be designed. However, the list must not be too long; if it is, then the eventual range of possible combinations of sub-solutions may become unmanageably large. About four to eight features or functions would make a sensible and manageable list.

For each feature or function list the means by which it might be achieved

These secondary lists are the individual sub-solutions which, when combined, one from each list, form the overall design solution. These sub-solutions can also be expressed in rather general terms, but it is probably better if they can be identified as actual components or physical embodiments. For instance, if one of the 'functions' of a vehicle is that it has motive power, then the different 'means' of achieving this might be engines using different fuels—e.g. petrol, diesel, electricity, gas.

The lists of means can include not only the existing, conventional components or sub-solutions of the particular product, but also new ones that you think might be feasible.

Draw up a chart containing all the possible sub-solutions

The morphological chart is constructed from the previous lists. At first, this is simply a grid of empty squares. Down the left-hand side is listed the essential features or functions of the product—i.e. the first list made earlier. Then across each row of the chart is entered the appropriate secondary lists of sub-solutions or means of achieving the functions. There is no relationship within the columns of the chart; the separate squares are simply convenient locations for the separate items. There might be, say, three means of achieving the first function, five means of achieving the second function, two means of achieving the third, and so on.

When it is finished, the morphological chart contains the complete range of all the theoretically possible different solution forms for the product. This complete range of solutions consists of the combinations made up by selecting one sub-solution at a time from each row. The total number of combinations is therefore often very large. For instance, if there were only three rows (functions) with three squares (means) in the first row, five in the second, and two in the third, then the complete set of possible combinations would number $3 \times 5 \times 2 = 30$. Because of this potential combinatorial explosion, the list of means for each function should be kept reasonably short.

Identify feasible combinations of sub-solutions

Clearly, for any product the complete range of possible combinations can be a very large number. Some of these combinations—probably a small number—will be existing solutions; some will be feasible new solutions; and some—possibly a great number—will be impossible solutions, for reasons of practically or because particular pairs of sub-solutions may be incompatible.

If the total number of possible combinations is not too large, then it may be possible to list each combination, and so to set out the complete range of solutions. Each potential solution can then be considered, and one or more of the better solutions (for reasons of cost, performance, novelty, or whatever criteria are important) chosen for further development.

If—as is more likely—the total number of possible combinations is very large, then some means has to be found of reducing this to something more manageable. One way of doing this is to choose only a restricted set of sub-solutions from each row—say, those that are known to be efficient or practical, or look promising for some other reason. Another way is to identify the non-feasible sub-solutions or incompatible pairs of sub-solutions, and so rule out those combinations that would include them.

A really exhaustive search of all the possible combinations in a morphological chart requires much patient and tedious work. (It is perhaps something where computer aids might help.) The only alternative is a more intuitive—or perhaps random—search of the chart for solutions.

SUMMARY

Morphological chart

Aim   To generate the complete range of alternative design solutions for a product, and hence to widen the search for potential new solutions.

Procedure
1. List the features or functions that are essential to the product. Whilst not being too long, the list must comprehensively cover the functions, at an appropriate level of generalization.
2. For each feature or function list the means by which it might be achieved. These lists might include new ideas as well as known existing components or sub-solutions.
3. Draw up a chart containing all the possible sub-solutions. This morphological chart represents the total solution space for the product, made up of the combinations of sub-solutions.
4. Identify feasible combinations of sub-solutions. The total number of possible combinations may be very large, and so search strategies may have to be guided by constraints or criteria.

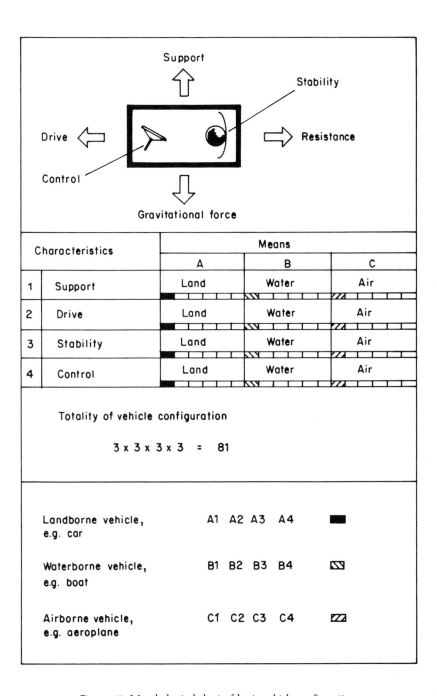

Figure 45. Morphological chart of basic vehicle configurations

EXAMPLES

Example 1: Vehicle configurations

A simple example of morphological analysis applied to vehicle configurations arose in the course of the design of vehicles for land and water speed record attempts, where it was necessary to define acceptable vehicle configurations. The functions of such vehicles were defined as:

Support
Drive
Stability
Control

In each case, the means of achieving these functions were classified as land, water or air. The resulting morphological chart is shown in Figure 45. You might like to try identifying other combinations than those defining a car, a boat or an aeroplane—e.g. a landyacht, a windsurf board, a hovercraft or a hydrofoil. (Source: Norris, 1963.)

Example 2: Potato harvesting machine

An improvement to the often rather abstract and wordy form of morphological charts can be made by using—where possible—pictorial representations of the different means for achieving the functions. An example is shown in Figure 46— a morphological chart for a potato harvesting machine. One selected combination of sub-solutions is highlighted in the chart. Notice that two sub-solutions are chosen for the sub-function 'separate stones'. This suggests either that each of these two sub-solutions is not really adequate on its own or else that the morphological chart itself has not really been constructed carefully enough— perhaps 'separate stones' is not just one sub-function, but needs to be more carefully defined. (Source: Pahl and Beitz, 1984.)

Example 3: Welding positioner

A welding positioner is a device for supporting and holding a workpiece and locating it in a suitable position for welding. Figure 47 shows a morphological chart for such a device, using words augmented by sketch diagrams.

One possible combination of sub-solutions is indicated by the zig-zag line through the chart. Even then, it was found that there were alternatives for the actual embodiment of some of the sub-solutions. For example, the sketches in Figure 48 show alternative configurations for the chosen means of enabling the tilting movement. (Source: Hubka, 1982.)

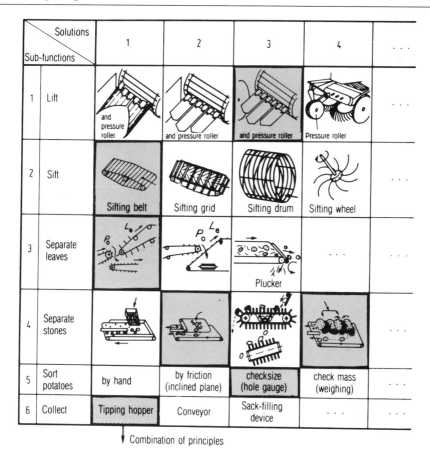

Figure 46. Morphological chart for a potato harvesting machine. The shaded boxes show one possible combination of sub-solutions

Example 4: Shaft coupling

This example shows that even small components can be usefully subject to morphological analysis. The example is that of a shaft coupling similar to the conventional 'Oldham' coupling, which transmits torque even in the case of radial and axial offsets of the shafts. Figure 49 shows a part of the morphological chart that was drawn up. One solution type (A) was analysed into its components and elements (presented here in columns, rather than rows) and the various sub-solutions listed in pictures and words. Two new alternatives combinations (B and C) are shown by the different sets of dots in the squares of the chart. One of these (B) was developed and patented as a novel design, as shown in Figure 50. (Source: Ehrlenspiel and John, 1987.)

Example 5: Field maintenance machine

This example shows an adaptation of the principles of morphological analysis. It

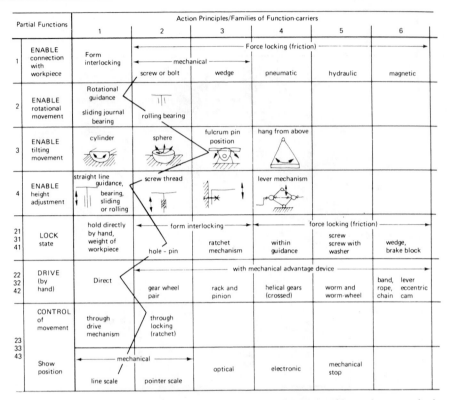

Partial Functions		Action Principles/Families of Function-carriers					
		1	2	3	4	5	6
1	ENABLE connection with workpiece	Form interlocking	← mechanical → screw or bolt	wedge	← Force locking (friction) → pneumatic	hydraulic	magnetic
2	ENABLE rotational movement	Rotational guidance, sliding journal bearing	rolling bearing				
3	ENABLE tilting movement	cylinder	sphere	fulcrum pin position	hang from above		
4	ENABLE height adjustment	straight line guidance, bearing, sliding or rolling	screw thread		lever mechanism		
21 31 41	LOCK state	hold directly by hand, weight of workpiece	← form interlocking → hole - pin	ratchet mechanism	← force locking (friction) → within guidance	screw, screw with washer	wedge, brake block
22 32 42	DRIVE (by hand)	Direct	← with mechanical advantage device → gear wheel pair	rack and pinion	helical gears (crossed)	worm and worm-wheel	band, rope, chain / lever, eccentric cam
23 33 43	CONTROL of movement	through drive mechanism	through locking (ratchet)				
	Show position	← mechanical → line scale	pointer scale	optical	electronic	mechanical stop	

Figure 47. Morphological chart for a welding positioner, with one possible combination of sub-solutions picked out by the zig-zag line

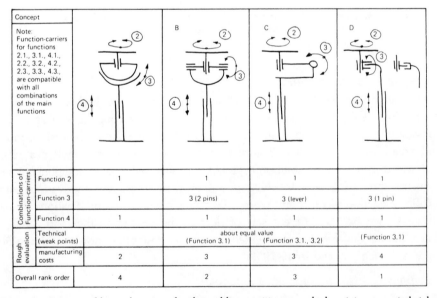

Concept			B	C	D
Note: Function-carriers for functions 2.1., 3.1., 4.1., 2.2., 3.2., 4.2., 2.3., 3.3., 4.3., are compatible with all combinations of the main functions					
Combinations of Function-carriers	Function 2	1	1	1	1
	Function 3	1	3 (2 pins)	3 (lever)	3 (1 pin)
	Function 4	1	1	1	1
Rough evaluation	Technical (weak points)		about equal value (Function 3.1)	(Function 3.1., 3.2)	(Function 3.1)
	manufacturing costs	2	3	3	4
Overall rank order		4	2	3	1

Figure 48. Four possible combinations for the welding positioner worked up into concept sketches

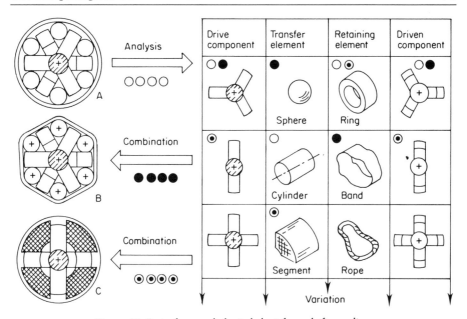

Figure 49. Part of a morphological chart for a shaft coupling

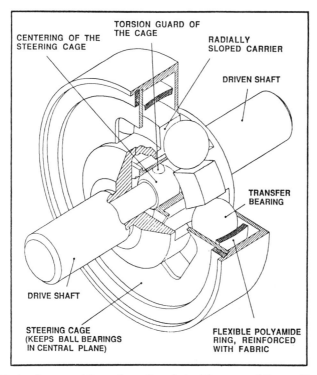

Figure 50. Design for a novel form of coupling, derived from one of the combinations in the morphological chart

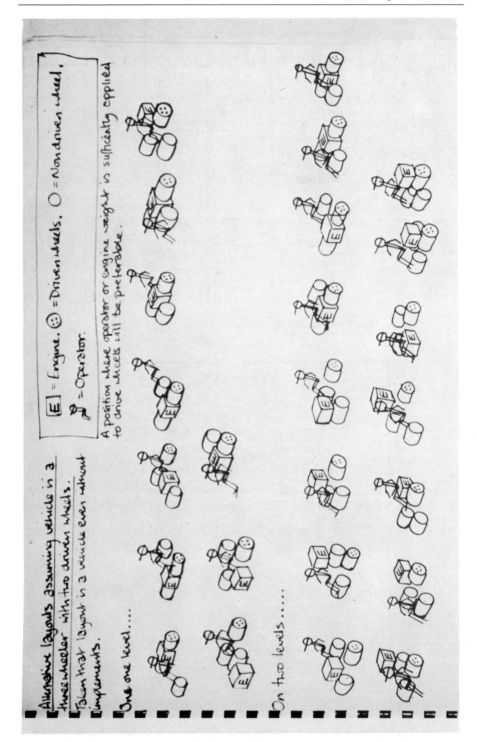

Alternative layouts assuming vehicle is a three wheeler with two driven wheels.

Taken that layout in a vehicle even without implements.

One one level.....

On two levels......

E = Engine. ☺ = Driven wheels. ○ = Non driven wheel.

🏃 = Operator.

A position where operator or engine weight is sufficiently applied to drive wheels will be preferable.

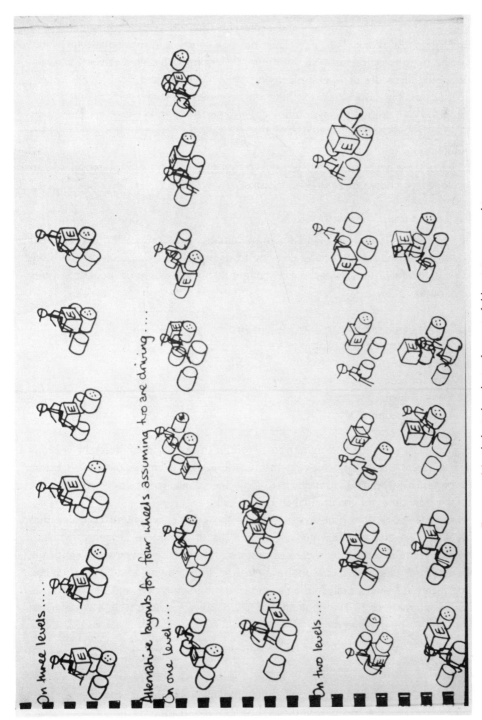

Figure 51. Morphological analysis of a sports field maintenance machine

is concerned solely with the form arrangement, or configuration, of the essential basic elements of the product, and represents the alternative configurations in purely graphical terms. The example is that of a sports field maintenance machine, and the morphological analysis in Figure 51 explores the alternatives for configuring the elements of operator, engine, driven wheels and non-driven wheels, and the possible disposition of these on one, two or three vertical levels. The sketches show alternative layouts for either a three- or four-wheel machine, with the arrangement options systematically varied so as to generate all possible design configurations.

The options were evaluated against design criteria, and one preferred option developed, as shown in Figure 52. (Source: Tovey, 1986.)

Worked example: Forklift truck

This example is concerned with finding alternative versions of the conventional forklift truck used for lifting and carrying loads in factories, warehouses, etc. If we investigate a few of these machines we might identify the essential generic features as follows:

1. Means of support which allows movement
2. Means of moving the vehicle
3. Means of steering the vehicle
4. Means of stopping the vehicle
5. Means of lifting loads
6. Location for operator

These features seem to be common to all forklift trucks, although different versions have different means of achieving the functions. For example, most such trucks run on wheels (means of support) that allow the vehicle to go anywhere on a flat surface, but some are constrained to run on rails.

When we look at the means of moving the vehicle, we might conclude that this is too general a feature and we decide that it should be broken down into separate features for (a) the means of propulsion (normally driven wheels), (b) the power source (such as electric motor, petrol or diesel engine) and (c) transmission type (gear and shafts, belt, hydraulic, etc.). Adding some new and perhaps rather fanciful alternatives to the conventional alternatives would enable a list like the following to be generated:

Feature	Means
Support	Wheels, track, air cushion, slides, pedipulators
Propulsion	Driven wheels, air thrust, moving cable, linear induction
Power	Electric, petrol, diesel, bottled gas, steam
Transmission	Gears and shafts, belts, chains, hydraulic, flexible cable

Figure 52. Design sketch for one preferred option derived from the morphological analysis of the maintenance machine

Steering	Turning wheels, air thrust, rails
Stopping	Brakes, reverse thrust, ratchet
Lifting	Hydraulic ram, rack and pinion, screw, chain or rope hoist
Operator	Seated at front, seated at rear, standing, walking, remote control

A morphological chart incorporating these lists is shown in Figure 53. You might like to calculate how many possible different solution combinations there are in this chart.

There are a staggering 90 000 possible forklift truck designs in the chart. Of course, some of these are not practicable solutions, or imply incompatible options; for example, an air cushion vehicle could not have steering by wheels. A typical, conventional forklift truck would comprise the following set of options from the chart:

Support	Wheels
Propulsion	Driven wheels
Power	Diesel engine
Transmission	Gears and shafts
Steering	Turning wheels

Feature	Means				
Support	Wheels	Track	Air cushion	Slides	Pedipulators
Propulsion	Driven wheels	Air thrust	Moving cable	Linear induction	
Power	Electric	Petrol	Diesel	Bottled gas	Steam
Transmission	Gears and shafts	Belts	Chains	Hydraulic	Flexible cable
Steering	Turning wheels	Air thrust	Rails		
Stopping	Brakes	Reverse thrust	Ratchet		
Lifting	Hydraulic ram	Rack and pinion	Screw	Chain or rope hoist	
Operator	Seated at front	Seated at rear	Standing	Walking	Remote control

Figure 53. Morphological chart for forklift trucks

Stopping   Brakes
Lifting   Rack and pinion
Operator   Seated at rear

The inclusion of a few unconventional options in the chart suggest some possibilities for radical new designs. For instance, the idea of 'pedipulators' (i.e. walking mechanisms similar to legs and feet) might lead to designs suitable for use on rough ground such as building sites—or even capable of ascending flights of steps!

The chart can also be used to help generate somewhat less fanciful but nonetheless novel design ideas. For example, the idea of using rails for steering might well be appropriate in some large warehouses, where the rails could be laid in the aisles between storage racks. The vehicle would have wheels for support and for providing propulsion. It would be electrically powered since it would be used indoors. One of the problems of electric vehicles is the limited battery power, so we might propose that our new design would pick up power from a live electric rail—like subway trains. This might be feasible in a fully automated warehouse which would not have the safety problems associated with people having to cross the rails. The 'operator' feature would therefore be remote control.

Feature	Means				
Support	Wheels	Track	Air cushion	Slides	Pedipulators
Propulsion	Driven wheels	Air thrust	Moving cable	Linear induction	
Power	Electric	Petrol	Diesel	Bottled gas	Steam
Transmission	Gears and shafts	Belts	Chains	Hydraulic	Flexible cable
Steering	Turning wheels	Air thrust	Rails		
Stopping	Brakes	Reverse thrust	Ratchet		
Lifting	Hydraulic ram	Rack and pinion	Screw	Chain or rope hoist	
Operator	Seated at front	Seated at rear	Standing	Walking	Remote control

Figure 54. One selected combination of sub-solutions from the morphological chart

A compatible set of sub-solutions for this new design therefore becomes:

Support	Wheels
Propulsion	Driven wheels
Power	Electric motor
Transmission	Belt
Steering	Rails
Stopping	Brakes
Lifting	Screw
Operator	Remote control

This set is shown as a selection from the morphological chart in Figure 54.

Chapter 8

EVALUATING ALTERNATIVES

When a range of alternative designs has been created, the designer is then faced with the problem of selecting the best one. At various points in the design process there may also be decisions of choice to be made between alternative sub-solutions or alternative features that might be incorporated into a final design. Choosing between alternatives is therefore a common feature of design activity.

Choices can be made by guesswork, by 'intuition', or by arbitrary decision. However, it is better if a choice can be made by some more rational, or at least open, procedure. Not only will the designer feel more secure in making the choice, but others involved in decision-making, such as clients, managers and colleagues in the design team, will be able to participate in or assess the validity of the choice.

If some of the previous design methods have already been used in a design process, then there should be some information available which should guide a choice between alternatives. For example, design proposals can be checked against criteria established by the performance specification method; and if design objectives have been established by the objectives tree method then these can be used in the evaluation of alternative designs.

In fact, the evaluation of alternatives can only be done by considering the objectives that the design is supposed to achieve. An evaluation assesses the overall 'value' or 'utility' of a particular design proposal with respect to the design objectives. However, different objectives may be regarded as having different 'values' in comparison with each other—i.e. may be regarded as being more important. Therefore it usually becomes necessary to have some means of differentially 'weighting' objectives, so that the performances of alternative designs can be assessed and compared across the whole set of objectives.

The *weighted objectives* method provides a means of assessing and comparing alternative designs, using differentially weighted objectives.

This method assigns numerical weights to objectives and numerical scores to the performances of alternative designs measured against these objectives. However, it must be emphasized that such weighting and scoring can lead the unwary into some very dubious arithmetic. Simply assigning numbers to objectives, or objects, does not mean that arithmetical operations can be applied to them. For instance, a football player assigned the number 9 is not necessarily three times as good as, or worth three times as much as a player assigned the number 3—even though he may score three times as many goals! Arithmetical operations can only be applied to data which have been measured on an *interval* or *ratio* scale.

The Weighted Objectives Method

PROCEDURE

List the design objectives

In order to make any kind of evaluation it is necessary to have a set of criteria, and these must be based on the design objectives—i.e. what it is that the design is meant to achieve. These objectives should have been established at an early point in the design process. However, at the later stages of the process—when evaluation becomes especially important—the early set of objectives may well have become modified, or may not be entirely appropriate to the designs that have actually been developed. Some clarification of the set of objectives may therefore be necessary as a preliminary stage in the evaluation procedure.

The objectives will include technical and economic factors, user requirements, safety requirements, and so on. A comprehensive list should be drawn up. Wherever possible, an objective should be stated in such a way that a quantitative assessment can be made of the performance achieved by a design on that objective. Some objectives will inevitably relate to qualitative aspects of the design; these may later be allocated 'scores', but the earlier warning about limitations on the use of arithmetic must be remembered.

Rank-order the list of objectives

The list of objectives will contain a wide variety of design requirements, some of which will be considered to be more important than others. As a first step towards determining relative 'weights' for the objectives, it is usually possible to list them in rank order of importance. One way of doing this is to write each objective on a separate card and then to sort the cards into a comparative rank order—i.e. from 'most important' to 'least important'.

As with many other aspects of this design method, it is usually helpful if the rank-ordering of objectives can be done as a team effort, since different members of a design team may well give different priorities to different objectives. Discussion of these differences will (hopefully!) lead to a team consensus. Alternatively, the client may be asked to decide the rank-ordering or market research might be able to provide customers' preferences.

The rank-ordering process can be helped by systematically comparing pairs of objectives, one against the other. A simple chart can be used to record the comparisons and to arrive at a rank order, like this:

Objectives	A	B	C	D	E	Row totals
A	—	0	0	0	1	1
B	1	—	1	1	1	4
C	1	0	—	1	1	3
D	1	0	0	—	1	2
E	0	0	0	0	—	0

Each objective is considered in turn against each of the others. A figure 1 or 0 is entered into the relevant matrix cell in the chart, depending on whether the first objective is considered more or less important than the second, and so on. For example, start with objective A and work along the chart row, asking 'is A more important than B?'. . .'than C?'. . .'than D?', etc. If it is considered more important, a 1 is entered in the matrix cell; if it is considered less important, a 0 is entered. In the example above, objective A is considered less important than all others except objective E.

As each row is completed, so the corresponding column can also be completed with an opposite set of figures; thus, if row A reads 0001 then column A must be 1110. If any pair of objectives is considered equally important, a ½ can be entered in both relevant squares.

When all pairs of comparisons have been made, the row totals indicate the rank order of objectives. The highest row total indicates the highest priority objective. In the example above, the rank order therefore emerges as:

B
C
D
A
E

It is here that one of the first problems of ranking may emerge, where relationships may not turn out to be transitive; i.e. objective A may be considered more important than objective B and objective B more important than objective C, but objective C may then be considered more important than objective A. Some hard decisions may have to be made to resolve such problems!

A rank-ordering is an example of an *ordinal* scale; arithmetical operations cannot be performed on an ordinal scale.

Assign relative weightings to the objectives

The next step is to assign a numerical value to each objective, representing its 'weight' relative to the other objectives. A simple way of doing this is to consider the rank-ordered list as though the objectives are placed in positions of relative importance, or value, on a scale of, say 1 to 10 or 1 to 100. In the example above, the rank-ordered objectives might be placed in relative positions on a scale of 1–10 like this:

```
10          B
9
8
7           C
6
5           D
4           A
3
2           E
1
```

The most important objective, B, has been given the value 10, and the others then given values relative to this. Thus, objective C is valued as about 70 per cent of the value of objective B; objective A is valued twice as highly as objective E; etc. The corresponding scale values are the relative weights of the objectives. (Note that the highest- and lowest-ranked objectives are *not* necessarily placed at the absolute top and bottom positions of the scale.)

If you can achieve such relative weightings, and feel confident about the relative positions of the objectives on the scale, then you have converted the ordinal rank-order scale into an *interval* value scale, which can be used for arithmetic operations.

An alternative procedure is to decide to share a certain number of 'points'— say, 100—amongst all the objectives, awarding points on relative value and making trade-offs and adjustments between the points awarded to different objectives until acceptable relative allocations are achieved. This can be done on a team basis, with members of the team each asked to allocate, or 'spend', a fixed number of total points between the objectives according to how highly they value them. If 100 points were allocated amongst objectives A to E in the earlier example, the results might be:

B	35
C	25
D	18
A	15
E	7

An objectives tree can be used to provide what is probably a more reliable method of assigning weights. The highest-level, overall objective is given the value 1.0; at each lower level, the sub-objectives are then given weights relative to each other but which also total 1.0. However, their 'true' weights are calculated as a fraction of the 'true' weight of the objective above them.

This is clarified by Figure 55. Each box in the tree is labelled with the objective's number (O_0, O_1, O_{11}, etc.) and given two values: its value relative to its neighbours at the same level and its 'true' value or value relative to the overall objective. Thus, in the example below, objectives O_2 and O_3 are regarded as of equal value, but each only half as valuable as objective O_1. Sub-objectives O_{11} and O_{12} are given values relative to each other of 0.67:0.33, but their 'true' values can only total 0.5 (the 'true' value of objective O_1) and are therefore calculated as $0.67 \times 0.5 = 0.34$ and $0.33 \times 0.5 = 0.16$.

Using this procedure it is easier to assign weights with some consistency because it is relatively easy to compare sub-objectives in small groups of two or three and with respect to a single higher-level objective. All the 'true' weights add up to 1.0, and this also ensures the arithmetical validity of the weights.

Establish performance parameters or utility scores for each of the objectives

It is necessary to convert the statements of objectives into parameters that can be measured, or at least estimated with some confidence. Thus, for instance, an objective for a machine to have 'high reliability' might be converted into a performance parameter of 'breakdowns per 10 000 hours running time', which might be either measured from available data or estimated from previous experience with that type of machine.

Some parameters will not be measurable in simple, quantifiable ways, but it may be possible to assign utility scores estimated on a points scale. The simplest scale usually has five grades, representing performance as follows:

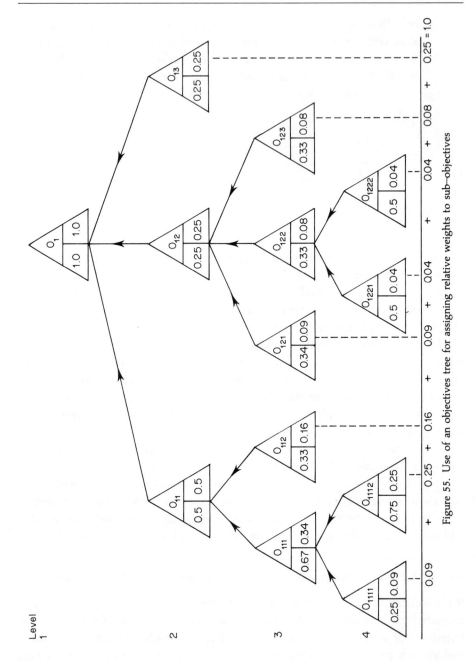

Figure 55. Use of an objectives tree for assigning relative weights to sub-objectives

Far below average
Below average
Average
Above average
Far above average

Often, a five-point scale (0–4) is too crude, and you will need to use perhaps a nine-point (0–8) or eleven-point (0–10) scale. The degrees of performance assessed by an eleven-point and a five-point scale might be compared as in Table 1.

TABLE 1

Eleven point scale	Meaning	Five point scale	Meaning
0	Totally useless solution		
		0	Inadequate
1	Inadequate solution		
2	Very poor solution		
		1	Weak
3	Poor solution		
4	Tolerable solution		
5	Adequate solution	2	Satisfactory
6	Satisfactory solution		
7	Good solution		
		3	Good
8	Very good solution		
9	Excellent solution		
		4	Excellent
10	Perfect or ideal solution		

Both quantitative and qualitative parameters can be compared together on a points scale, representing the worst-to-best possible performance range. For example, the fuel consumption and, say, the comfort of a motorcar could be represented on a seven-point scale as in Table 2.

TABLE 2

Points	Fuel consumption (miles/gal)	Comfort
0	< 27	Very uncomfortable
1	29	Poor comfort
2	32	Below average comfort
3	35	Average comfort
4	38	Above average comfort
5	41	Good comfort
6	> 43	Extremely comfortable

Care must be taken in compiling such points scales, because the values ascribed to the parameters may not rise and fall linearly. For example, on the scale above, the value of decreasing fuel consumption is assumed to increase linearly, but it might well be regarded as more valuable to provide improvements in fuel consumption at the lower end of the scale rather than the upper end; i.e. the 'utility curve' for a parameter might be an exponential or other curve, rather than linear.

Calculate and compare the relative utility values of the alternative designs
The final step in the evaluation is to consider each alternative design proposal and to calculate for each one a score for its performance on the established

parameters. Once again, the participation of all members of the design team is recommended (and especially those whose views ultimately count, such as customers!), since different solutions may be scored differently by different people.

The raw performance measures or points scores on each parameter for each alternative design must be adjusted to take account of the different weights of each objective. This is done by simply multiplying the score by the weight value, giving a set of adjusted scores for each alternative design which indicates the relative 'utility value' of that alternative for each objective.

These utility values are then used as a basis of comparison between the alternative designs. One of the simplest comparisons that can be made is to add up the utility value scores for each alternative. These total scores then allow the alternatives to be ranked in order of overall performance.

Other comparisons are possible, such as drawing graphs or histograms to represent the utility value 'profiles' of the alternative designs. These visual, rather than numerical, comparisons present a 'picture' which may be easier to absorb and reflect on. They also highlight where alternatives may be significantly different from each other in their performance.

The benefit of using this evaluation method often lies in making such comparisons between alternatives, rather than using it simply to try to choose the 'best' alternative. Many rather contentious weightings, points scores and other decisions will probably have been made in compiling the evaluation, and some of the arithmetic may well be highly dubious. The 'best' overall utility value may therefore be highly misleading, but the discussions, decisions, rankings and comparisons involved in the evaluation are certain to have been illuminating.

SUMMARY

Weighted objectives

Aim   To compare the utility values of alternative design proposals, on the basis of performance against differentially weighted objectives.

Procedure
1. List the design objectives. These may need modification from an initial list; an objectives tree can also be a useful feature of this method.
2. Rank-order the list of objectives. Pair-wise comparisons may help to establish the rank order.
3. Assign relative weightings to the objectives. These numerical values should be on an interval scale; an alternative is to assign relative weights at different levels of an objectives tree, so that all weights sum to 1.0.

4. Establish performance parameters or utility scores for each of the objectives. Both quantitative and qualitative objectives should be reduced to performance on simple points scales.
5. Calculate and compare the relative utility values of the alternative designs. Multiply each parameter score by its weighted value—the 'best' alternative has the highest sum value. Comparison and discussion of utility value profiles may be a better design aid than simply choosing the 'best'.

EXAMPLES

Example 1: Motorcar

The basic principles of the weighted objectives method are demonstrated in this simple example of the calculation of utility values for three different motorcars. Given that all three are about the same purchase price, the potential purchaser might set the following objectives:

Low fuel consumption
Low cost of spare parts
Easy to maintain
High comfort

These are considered to have the respective relative weights of 0.5, 0.2, 0.1 and 0.2.

The objective of 'low fuel consumption' is easily assigned a performance parameter—that of the known figure for fuel consumption from standard tests, in miles per gallon. The figures are converted in the comparison table (Figure 56) to points on a seven-point scale, as suggested in the Procedure.

To measure 'cost of spare parts' it was decided to take the cost of a standard set of typical parts that might have to be purchased in, say, the first three years of a car's life. Again, a seven-point scale was used to convert the cost figures to a utility points score.

The objective of 'easy to maintain' was assessed on a seven-point scale of 'simplicity of servicing'. The assessment was based on the number of routine service tasks and the ease with which these could be done for each particular motorcar.

Finally, comfort was also assessed on a seven-point scale, similar to that presented in the Procedure.

For each alternative, the utility score for each objective is multiplied by the objective's weight, giving a relative utility value. If these are added together, an overall utility value for each alternative is obtained. As the figures in the chart of Figure 56 show, car B emerges as clearly the 'best' overall.

Objective	Weight	Parameter	Car A			Car B			Car C		
			Magnitude	Score	Value	Magnitude	Score	Value	Magnitude	Score	Value
Low fuel consumption	0.5	Miles per gallon	33	2	1.0	40	4	2.0	36	3	1.5
Low cost of spare parts	0.2	Cost of 5 typical parts	£18	7	1.4	£22	5	1.0	£28	2	0.4
Easy to maintain	0.1	Simplicity of servicing	Very simple	5	0.5	Com-plicated	2	0.2	Average	3	0.3
High comfort	0.2	Comfort rating	Poor	2	0.4	Very good	5	1.0	Good	4	0.8
Overall utility value					3.3			4.2			3.0

Figure 56. Weighted objectives evaluation chart for three alternative motorcars

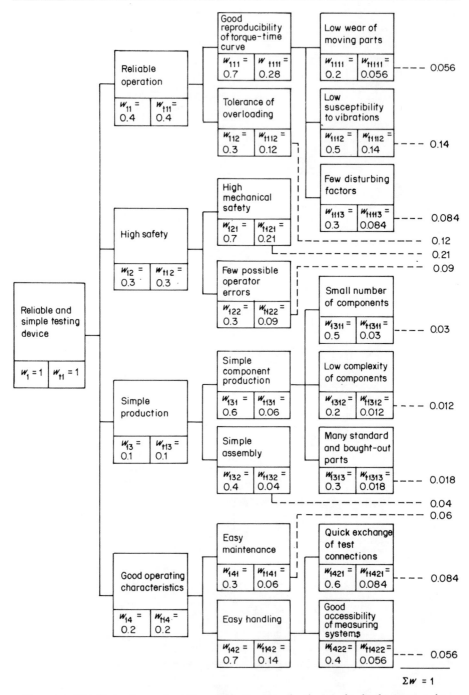

Figure 57. Weighted values assigned to an objectives tree for the impulse—load testing machine

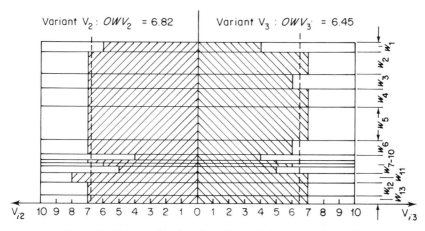

Figure 59. Value profiles for alternative test rig designs V_2 and V_3

Example 2: Test rig

This example is taken from the project for the design of a laboratory rig for carrying out impulse–load tests on shaft connectors. A thorough evaluation was made of a number of alternative designs, based on the objectives tree which was presented earlier in the objectives tree method.

The objectives and sub-objectives at different levels were weighted in the manner described in the Procedure (Figure 57). The design team then went on to devise measurable or assessable parameters for all of the objectives, as indicated in the comparison chart.

Utility values were calculated for each objective, for each of four alternative designs (Figure 58). The second alternative (variant V_2) emerges as the 'best' solution, with an overall utility value of 6.816. However, variant V_3 seems quite comparable, with an overall utility value of 6.446. A comparison of the 'value profiles' of these two alternatives was therefore made. This is shown in Figure 59, where the thickness of each bar in the chart represents the relative weight of each objective and its length represents the score for that objective achieved by the particular design. The chart shows that V_2 has a more consistent profile than V_3, with fewer relatively weak spots in its profile. V_2 therefore seems to be a good, 'all-round' design, and the comparison confirms it as being the best of the alternatives. However, improvement of V_3 in perhaps just one or two of its lower-scoring parameters might easily push it into the lead. (Source: Pahl and Beitz, 1984.)

Example 3: Swivel joint

This problem was the design of a swivel joint used in an underwater marine environment as part of a current-metering system. A previous design was considered unsuitable because of its high cost and poor performance (especially the high friction between adjacent moving parts).

No.	Evaluation criteria	Parameters	Unit	Wt.	Variant V_1 Magn. m_{i1}	Value V_{i1}	Weighted value WV_{i1}	Variant V_2 Magn. m_{i2}	Value V_{i2}	Weighted value WV_{i2}	Variant V_3 Magn. m_{i3}	Value V_{i3}	Weighted value WV_{i3}	Variant V_4 Magn. m_{i4}	Value V_{i4}	Weighted value WV_{i4}
1	Low wear of moving parts	Amount of wear	-	0.056	high	3	0.168	low	6	0.336	average	4	0.224	low	6	0.336
2	Low susceptibility to vibrations	Natural frequency	s^{-1}	0.14	410	3	0.420	2370	7	0.980	2370	7	0.980	<410	2	0.280
3	Few disturbing factors	Disturbing factors	-	0.084	high	2	0.168	low	7	0.588	low	6	0.504	(average)	4	0.336
4	Tolerance of overloading	Overload reserve	%	0.12	5	5	0.600	10	7	0.840	10	7	0.840	20	8	0.960
5	High mechanical safety	Expected mechan. safety	-	0.21	average	4	0.840	high	7	1.470	high	7	1.470	very high	8	1.680
6	Few possible operator errors	Possibilities of operator errors	-	0.09	high	3	0.270	low	7	0.630	low	6	0.540	average	4	0.360
7	Small number of components	No. of components	-	0.03	average	5	0.150	average	4	0.120	average	4	0.120	low	6	0.180
8	Low complexity of components	Complexity of components	-	0.012	low	6	0.072	low	7	0.084	average	5	0.060	high	3	0.036
9	Many standard and bought-out parts	Proportion of standard and bought-out components	-	0.018	low	2	0.036	average	6	0.108	average	6	0.108	high	8	0.144
10	Simple assembly	Simplicity of assembly	-	0.04	low	3	0.120	average	5	0.200	average	5	0.200	high	7	0.280
11	Easy maintenance	Time and cost or maintenance	-	0.06	average	4	0.240	low	8	0.480	low	7	0.420	high	3	0.180
12	Quick exchange of test connections	Estimated time needed to exchange test connections	min	0.084	180	4	0.336	120	7	0.588	120	7	0.588	180	4	0.336
13	Good accessibility of measuring systems	Accessibility of measuring systems	-	0.056	good	7	0.392	good	7	0.392	good	7	0.392	average	5	0.280
				$\Sigma w_i = 1.0$		$OV_1 = 51$ $R_1 = 0.39$	$OWV_1 = 3.812$ $WR_1 = 0.38$		$OV_2 = 85$ $R_2 = 0.65$	$OWV_2 = 6.816$ $WR_2 = 0.68$		$OV_3 = 78$ $R_3 = 0.60$	$OWV_3 = 6.446$ $WR_3 = 0.64$		$OV_4 = 68$ $R_4 = 0.52$	$OWV_4 = 5.388$ $WR_4 = 0.54$

Figure 58. Completed evaluation chart for four alternative designs for the impulse–load testing machine

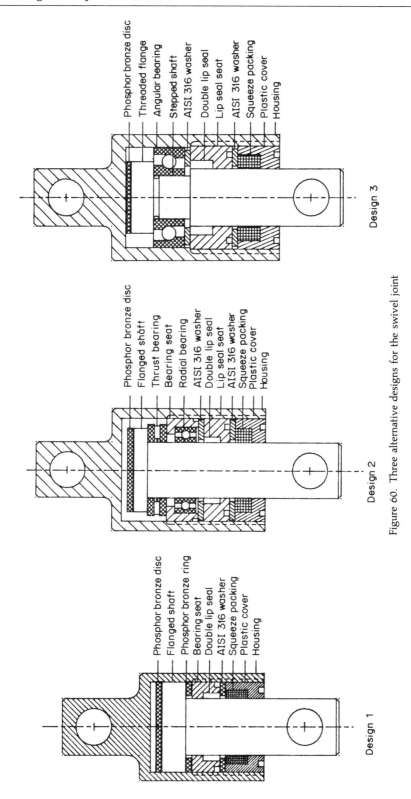

Figure 60. Three alternative designs for the swivel joint

Three different new designs were developed and evaluated by weighted objectives (Figure 60). Figure 61 shows the evaluation chart; each design is scored on a 0–10 scale for each objective, and each score multiplied by a weighting factor to give a utility value. Because of the nature of the problem, cost factors were given relatively high weights, whilst strength was low-weighted. This was because the operational loads were not severe and the materials were chosen more for resistance to corrosion than for loading stresses.

Designs 1 and 3 achieved similar overall utility values, but design 1 was considered less complex in manufacture and was therefore chosen in preference to design 3. (Source: Shahin, 1988.)

Example 4: City car

This example continues the problem of designing a small car for use in the city or over a very limited journey range, introduced in the performance specification method (Example 4). As a part of the design study, the designers drew up a morphological chart of six basic types of city car and the variants within each type for aspects such as the positioning of the engine. This total set of variants

Design criteria	Weight	Design 1		Design 2		Design 3	
	W^*	S	U	S	U	S	U
1 Cost							
Materials	6	8.5	0.51	5.5	0.33	7	0.42
Seals	2	8	0.16	8	0.16	8	0.16
Bearings	4	9	0.36	5	0.2	8	0.28
Washers	1	7.5	0.07	7.5	0.07	7.5	0.07
Squeeze packing	2	9	0.18	9	0.18	9	0.18
Bolts	1	9	0.09	9	0.09	8	0.08
Labour	6	8	0.48	5	0.3	7.5	0.45
Tools and equipment	6	8	0.48	5	0.3	7.5	0.45
Indirect cost	20	8.5	1.7	7	1.4	7.5	1.5
Marketing	2	7	0.14	8	0.16	9	0.18
2 Performance							
Sealing	9	8	0.72	8	0.72	8	0.72
Smoothness	9	5	0.45	9	0.81	8.5	0.76
Alignment	6	5	0.3	7	0.42	8	0.48
Growth formation	2	8	0.16	8	0.16	8	0.16
Maintenance	4	8	0.32	8	0.32	8	0.32
3 Manufacturing							
Ease	5	8.5	0.42	7	0.35	7.5	0.37
Time	5	9	0.45	4.5	0.22	7.5	0.37
Assembly	5	9	0.45	6.5	0.32	8	0.4
4 Strength	5	8	0.4	9.5	0.47	9.5	0.47
The overall utility			7.84		6.98		7.82

$^*W$ = percentage weight of each criterion (from 100)
S = score of quality of each design (from 10)
U = utility (weighted score) of design = $W \times S$

Figure 61. Evaluation chart for the three designs

was then evaluated, using weighting factors and an evaluation of each variant on a scale of 0–10 (Figure 62). From this, car type 4 emerges as the preferred basic form, and was used as the concept for more detailed design work. (Source: Pighini *et al.*, 1983.)

Worked example: Drive unit

This example of the weighted objectives evaluation method uses estimates of money costs rather than the points-scale system. The example arose in the evaluation of five alternative drive units for a machine that processes synthetic fibre (Source: Tebay *et al.*, 1984). The objectives used in the evaluation were:

1. Cost
2. Standardization with other units on plant
3. Additional features offered as standard
4. Reliability

A monetary value for 'cost' is, of course, immediately available, but is not so obvious for the other objectives. For the second objective, therefore, estimates were made of costs of additional spares required if a non-standardized unit were chosen. For the third objective it was possible to estimate the cost of incorporating the required additional features, on units for which they were not standard.

Estimating a monetary value for reliability was more difficult. It was decided as a first step to assign subjective relative values to each variant, reflecting the designer's degree of confidence in the unit (based on experience), on a scale from 1 (absolute confidence) to 0 (no confidence). This resulted in Table 3.

TABLE 3

	V_1	V_2	V_3	V_4	V_5
Probability of no problem	1	0.8	1	1	0.9
Probability of some problems	0	0.2	0	0	0.1

The cheapest drive unit with a confidence rating of 1 was V_1 (cost: £355). If problems occurred with units V_2 or V_5, they could be replaced with a V_1 unit, giving relative costs (allowing £50 freight costs in addition to the basic costs of a V_1 unit) as shown in Table 4.

	Sub-function	Function carrier	Task	Principle of evaluation	Weight factor	Type 1	
A	Offers space, support, protection for people and luggage	Body and frame of car	Optimal form for car	Internal space Protective ability	0.12	5 	
B	Generates the power for transmission	Motor and transmission	Optimal position for motor	Available space Complexity of transmission	0.08	7 	8 
C	Supports people in a safe and comfortable way	Seats	Optimal disposition for seats	Safety Comfort Possibility of getting 4 seats	0.08	4 	5 
D	Offers space for luggage	Luggage room	Optimal position and higher capacity of luggage room	Space used for luggage Internal room	0.08	5 	4 
E₁ E₂	For entry and exit For views outside	Doors Windows	Optimal number, dimensions and position of doors and windows	Facilities for entry, exit and putting in luggage Visibility for driver and passengers	0.08	7 	7 
F	Changes direction of car driving	Steering system	Optimal disposition of steering system	Position of motor Complexity of steering system	0.08	9 	10 
G	Aesthetic evaluation				0.08	9	
H	Cost evaluation				0.16	4	
I	Safety evaluation				0.24	4	5
	Total sum					5.48	5.88
	Order of merit					9	7

NOTES: 1 Evaluation marks 0 = unacceptable; 1-3 = still acceptable; 4-6 = fair;
7-9 = good; 10 = very good (optimal solution)
2 Total sum = (mark × weighting factor)
3 *The marks in parentheses are for the car with two seats

Figure 62. Evaluation chart for alternative concepts of the small city car

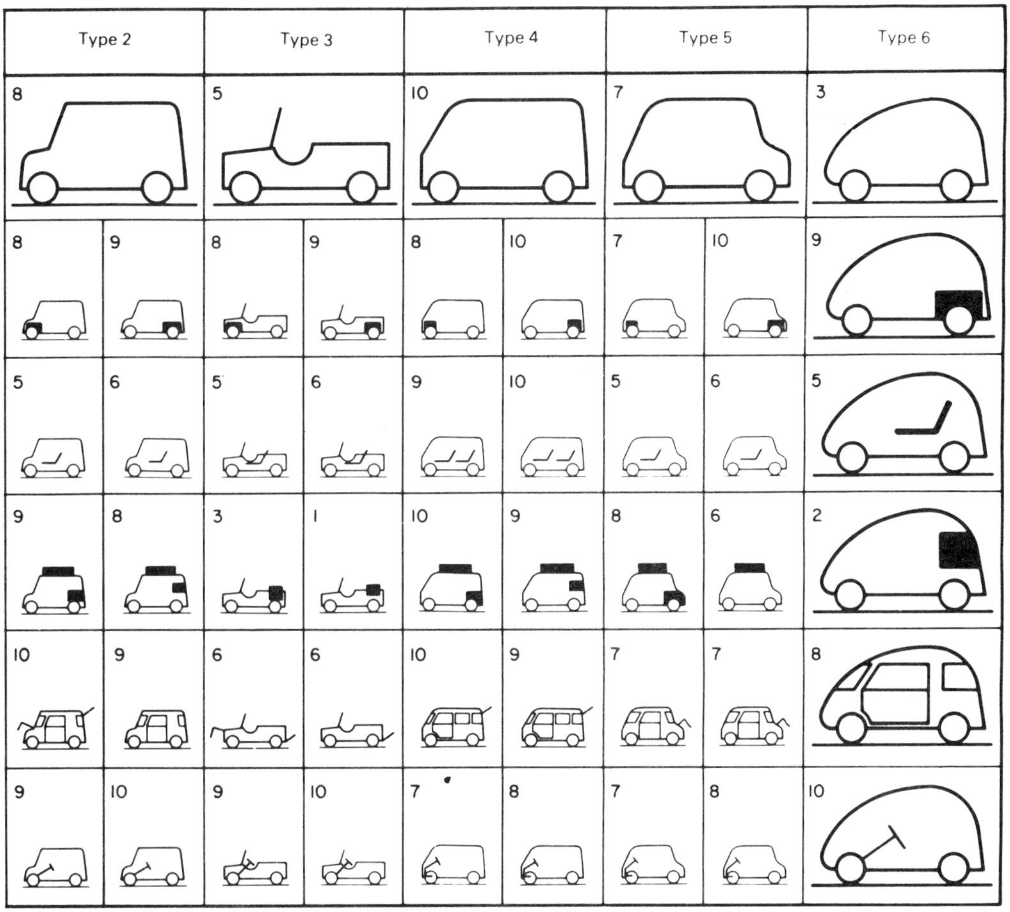

Figure 62 (*cont.*).

TABLE 4

	V_1	V_2	V_3	V_4	V_5
Reliability cost	0	405 ×0.2 —— £81	0	0	405 ×0.1 —— £40.5

The objectives were not weighted in any sophisticated way, but were simply ranked in order of importance and given a weight equivalent to their rank order, as follows:

	Rank order	Weight
Objective 1	3	2
Objective 2	1	4
Objective 3	2	3
Objective 4	4	1

This resulted in Table 5, a table of costs.

TABLE 5

Variant	Objective	1	2	3	4	Weighted
	Weight	2	4	3	1	total cost
V_1	Cost	335	0	50	0	
	Weighted cost	710	0	150	0	860
V_2	Cost	300	40	50	81	
	Weighted cost	600	160	150	81	991
V_3	Cost	435	0	0	0	
	Weighted cost	870	0	0	0	870
V_4	Cost	385	20	0	0	
	Weighted cost	770	80	0	0	850
V_5	Cost	485	60	25	40.5	
	Weighted cost	970	240	75	40.5	1325.5

The figures for weighted total cost suggest that V_4 is the 'best' alternative, having the lowest cost. However, the designer actually preferred to choose V_3.

This suggests that the initial weighting of objectives was incorrect. For V_3 to be preferred implies that objective 1 (the basic cost of the unit) is not assigned a weight of 2, but a weight of 1; i.e. the ranking and weighting of objectives should have been:

	Rank order	Weight
Objective 1	4	1
Objective 2	1	4
Objective 3	2	3
Objective 4	3	2

Given these alternative weights, the weighted total costs for the five variants can be calculated as:

V_1	505
V_2	772
V_3	435
V_4	465
V_5	881

Thus, with these revised weightings, V_3 is the lowest cost alternative. This example shows that the process of calculating weighted utility values helps to clarify preferences and priorities that might otherwise remain obscure. When they are out in the open they can at least be discussed and justified, and perhaps revised. (Source: Tebay, Atherton and Wearne, 1984.)

Chapter 9

IMPROVING DETAILS

A great deal of design work in practice is concerned not with the creation of radical new design concepts but with the making of modifications to existing product designs. These modifications seek to improve a product—to improve its performance, to reduce its weight, to lower its cost, to enhance its appearance, and so on. All such modifications can usually be classified into one of two types: they are either aimed at increasing its value to the purchaser or at reducing its cost to the producer.

The *value* of a product to its purchaser is what he or she thinks the product is worth. The *cost* of a product to its producer is what it costs to manufacture and deliver it to the point of sale. A product's selling *price* normally falls somewhere between its cost to the producer and its value to the purchaser.

Designing is therefore essentially concerned with adding value. When raw materials are converted into a product, value is added over and above the basic costs of materials and their processing. How much value is added depends on the perceived worth of the product to its purchaser, and the perception is substantially determined by the attributes of the products as provided by the designer.

Of course, values fluctuate, depending on social, cultural, technological and environmental contexts, which change the need for, relevance or usefulness of a product. There are also complex psychological and sociological factors which affect the symbolic or esteem value of a product, but there are also more stable and comprehensible values associated with a product's function, and it is principally these functional values that are of concern to the engineering designer.

The *value engineering* method focuses on functional values, and aims to increase the difference between the cost and value of a product: by lowering cost or adding value, or both. In many cases, the emphasis is simply on reducing costs, and the design effort is concentrated onto the detailed design of components, on their materials, shapes, manufacturing methods and assembly processes. This more limited version of the method is known as *value analysis*. It is usually applied only to the refinement of an existing product, whereas the broader value engineering method is also applicable to new designs or to the substantial redesign of a product. Value analysis particularly requires detailed information on component costs.

Because of the variety and detail of information required in value analysis and value engineering, they are usually conducted as team efforts, involving members from different departments of a company, such as design, costing, marketing, production departments, etc.

The Value Engineering Method

PROCEDURE

List the separate components of the product and identify the function served by each component

One of the ways in which companies seek to better their rivals' products is to buy an example of the competing product, strip it down to its individual components and try to learn how their own product might be improved in both design and manufacture. This is one way of learning some of a competitor's secrets without resorting to industrial espionage.

The same sort of technique is at the heart of value engineering and value analysis. The first analytical step in the method is to strip a product down to its separate components—either literally and physically or by producing an itemized parts list and drawings. However, parts lists and conventional engineering drawings are of limited value in understanding and visualizing the components, the ways in which they fit together in the product overall and how they are manufactured and assembled. Therefore, if an actual product, or a prototype version, is not available for dismantling, then something like exploded diagrams of the product are helpful in showing components in three-dimensional form and in their relative locations or assembly sequences.

The purpose of this first step in the procedure is to develop a thorough familiarity with the product, its components and their assembly. This is particularly important if a team is working on the project, since different team members will have different views of the product, and perhaps only limited understandings of the components and their functions. It is therefore necessary to go through an exhaustive analysis of the sub-assemblies and individual components, and how they contribute in functional terms to the overall product.

Sometimes it is not at all clear what function a component serves or contributes to! This may be found particularly in products that have had a long life and may have gone through many different versions: some components may simply be redundant items left over from earlier versions. However, it may also be the case that components have been introduced to cope with problems that arose in the use of the product, and so any components that may appear to be redundant should not be dismissed too readily. Sometimes redundancy is even deliberately designed in to a product in order to improve its reliability.

The objective of this step in the procedure is to produce a complete list of components, grouped as necessary into sub-assemblies with their identified functions. In value engineering, rather than the more limited value analysis applications of this method, a similar objective applies, even though the ultimate intention might be to develop a completely new product, rather than just to make improvements to an existing product. In this case, the starting point might be an existing product, against which it has been decided to compete in the market, or an 'archetypal' or hypothetically typical version of the proposed new product.

Determine the values of the identified functions

Questions of value are, of course, notoriously difficult. They are the stuff of political debate and of subjective argument between individuals. Reaching agreement in a team on the 'value' of particular product functions therefore may not be easy. However, it must be remembered that the value of a product means its value as perceived by its purchaser. Thus the values of product functions must be those as perceived by customers, rather than by designers or manufacturers. Market research must therefore be the basis of any reliable assessment of the values of functions.

The market prices of different products can sometimes provide indicators of the values that customers ascribe to various functions. For instance, some products exist in a range of different versions, with more functions being incorporated in the products at the higher end of the range. Differences in prices should therefore reflect differences in the perceived values of the additional functions. However, customers are likely to 'perceive' a product as a total entity, rather than as a collection of separate functions, and subjective factors such as appearance are often of more importance than objective functional factors. It is said that the solidity of the 'clunk' made by closing a car door is one of the most important factors influencing a customer's perception of the value of a motorcar!

Considerable efforts have been put into trying to quantify perceived values or benefits, particularly in connection with the 'cost–benefit analysis' method used in planning. For example, in transport planning, some of the benefits of a new road or bridge can be quantified in terms of the time saved by travellers in using the new facility. Attempts are then made to convert all such benefits (and costs) into monetary terms, so that direct comparisons can be made.

Despite the difficulty of assessing values, it is necessary to make the best attempt one can to rationalize and express the perceived values of component functions. It can be pointless to reduce the costs of components if their values are also being reduced, so that the product becomes less desirable (or 'valuable') to prospective purchasers. If quantified and reliable estimates of values cannot be made, then at least simple assessments of high/medium/low value can be attempted.

Determine the costs of the component

Surprising though it may seem, it is not always easy for a company to determine the exact costs of components used in products. The company's accounting methods may not be sufficiently specific for itemized component costs to be identifiable. One of the useful by-products of a value analysis or value engineering exercise, therefore, can be the improvement of costing methods. Team working in the exercise again becomes particularly relevant, because reliable cost information at sufficient detail may only be obtainable by synthesizing information from different departmental specialists.

It is not sufficient to know just the cost of the material in a component, or even its bought-in cost if it is obtained from a supplier. The value analysis team needs to know the cost of the component as an element of the overall product cost—i.e. after it is fully finished and assembled into the product. Therefore, as well as material or bought-in costs there are labour and machine costs to be added for the assembly processes. It is sometimes suggested that factory overhead costs should also be added, but these can be very difficult to assign accurately to individual components, and instead can perhaps be assumed to be spread equally over all components.

It is important not to ignore low-cost components, particularly if they are used in large numbers (e.g. screws or other fasteners). Even a relatively small cost reduction per item can amount to a substantial overall saving when multiplied by the number of components used.

As well as determining the absolute costs of components, their relative or percentage costs in terms of the total product cost should also be calculated. Attention might then be focused on components or sub-assemblies which represent a significant portion of the total cost.

Search for ways of reducing cost without reducing value or of adding value without adding cost

This fundamental design stage calls for a combination of both critical and creative thinking. The critical thinking is aimed at what the design *is* and the creative thinking is aimed at what it *might be*. The concept of stripping down a competitor's product to look for ways of improving on it is a useful one to bear in mind at this stage. It is usually easier to be critical of and to suggest improvements to someone else's design rather than one's own, and it is this kind of creative criticism that is needed at this final stage.

Attempts to reduce costs usually focus on components and on ways of simplifying their design, manufacture or assembly, but the functions performed by a product should also be looked at critically, because it may be possible to simplify them, reduce their range or even eliminate them altogether if they are of limited value to the purchaser.

There are some general strategies which can be applied in order to direct the search for ways or reducing costs. The first is to concentrate on high-cost components, with a view to substituting lower-cost alternatives. The second is to review any components used in large numbers, since small individual savings may add up to a substantial overall saving. A third strategy is to identify components and functions which are matched as high-cost/high-value or low-cost/low-value, since the aim is to achieve high-value functions with low-cost components. One particular technique is to compare the cost of a component used in the design with the absolute lowest-cost means of achieving the same function; large differences suggest areas for cost reduction, even though the lowest-cost version may not be a viable option.

A checklist of cost-reduction guidelines can be followed:

Eliminate	Can any function, and therefore its components, be eliminated altogether? Are any components redundant?
Reduce	Can the number of components be reduced? Can several components be combined into one?
Simplify	Is there a simpler alternative? Is there an easier assembly sequence? Is there a simpler shape?
Modify	Is there a satisfactory cheaper material? Can the method of manufacture be improved?
Standardize	Can parts be standards rather than specials? Can dimensions be standardized or modularized? Can components be duplicated?

Whilst the value analysis approach tends to emphasize reducing costs, the broader value engineering approach looks also for ways of adding value to a product. For example, rather than eliminating functions, as suggested above, value engineering might seek ways of improving or enhancing a product's functions. Nevertheless, the aim is always to increase the value/cost ratio.

One of the most significant means of adding value to a product, without necessarily increasing its cost, is to improve its ease of use. This has become particularly evident with the preference for personal computers which are found to be 'user friendly'. In this case, the 'friendliness' perhaps applies more to the computer's software than its hardware, or at least to the combination of software and hardware such that use of the computer seems natural and easy. However, similar principles can be applied to all machines—their use should be straightforward, clear and comfortable. There is a considerable body of knowledge in the field of ergonomics which can be applied to these user aspects of machine design.

Other attributes which commonly contribute to the quality or value of a product are:

Utility	Performance on aspects such as capacity, power, speed, accuracy or versatility
Reliability	Freedom from breakdown or malfunction; performance under varying environmental conditions
Safety	Secure, hazard-free operation
Maintenance	Simple, infrequent or no maintenance requirements
Lifetime	Except for disposable products, a long lifetime which offers good value for the initial purchase price
Pollution	Little or no unpleasant or unwanted by-products, including noise and heat

Finally, there is a whole class of value attributes related to aesthetics. This includes not only the appearance of a product—colour, form, style, etc.—but also aspects such as surface finish and feel to the touch.

Evaluate alternatives and select improvements

The application of value analysis or value engineering should result in a number of alternative suggestions for changes to the product design. Some of these alternatives might well be incompatible with each other, and in fact all suggestions should be carefully evaluated before selecting those that can be shown to be genuine improvements.

SUMMARY

Value engineering

Aim To increase or maintain the value of a product to its purchaser whilst reducing its cost to its producer.

Procedure

1. List the separate components of the product and identify the function served by each component. If possible, the actual product should be disassembled into its components; exploded diagrams and component-function charts are more useful than parts lists.
2. Determine the values of the identified functions. These must be the values as perceived by customers.
3. Determine the costs of the components. These must be after fully finished and assembled.
4. Search for ways of reducing cost without reducing value or of adding value without adding cost. A creative criticism is necessary, aimed at increasing the value/cost ratio.
5. Evaluate alternatives and select improvements.

EXAMPLES

Example 1: Turbine gears

In value analysis it is useful to understand not only the actual costs of components but also their relative costs as proportions of the overall product cost. Figure 63 shows two types of 'cost structure' analysis for a set of turbine gears. The first table gives the actual cost of each component and its relative cost as a percentage of the overall manufacturing cost of the product. The second table shows the percentages of different types of cost that are involved in making each component. These data help the designer to understand the proportional costs of the product components, and hence to apply cost-reduction measures where they will be most effective—in this case, for example, the pinion shaft rather than the pinion bearings, or the casing rather than the seals and covers. (Source: Ehrlenspiel, 1987.)

Example 2: Generator

It is often helpful to have cost structure data presented graphically, as in Figure 64 for a synchronous generator. The use of such a visual aid makes it easier to develop an intuitive understanding of proportional costs, as well as clearly identifying the areas where cost-reduction measures might be most effective. In this example, it is clear that it is hardly possible to reduce labour and overhead costs for some components, such as the rotor shaft (component R_1), and that material costs might therefore be looked into. However, both labour and overhead costs look like candidates for reduction in the rotor shaft (R_2), and all three—materials, labour and overheads—for the stator housing (S_3). (Source: Pahl and Beitz, 1984)

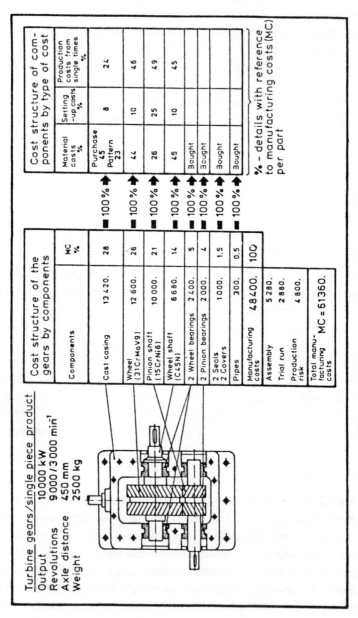

Figure 63. Cost structure of a set of turbine gears

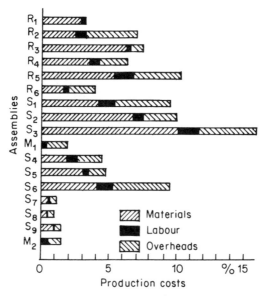

Figure 64. Component costs for a generator

Parts or assemblies	Cost (£)
Banjo assembly	1.07
Valve body	6.62
Spring	0.39
Diaphragm assembly	2.14
Cover	2.24
Lug	0.10
Nuts, bolts and washers	2.34
Assembly cost	4.58
Total	£19.48

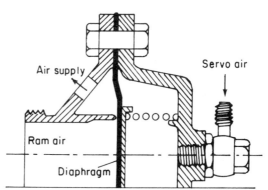

Figure 65. Cost analysis of an aircraft air valve

Example 3: Air valve

This example (Figure 65) shows how components and functions can be costed in a comparison table or matrix. Components often contribute to several different or related functions, and hence the cost of a particular function is often spread across several components. The kind of component/function cost matrix shown in Figure 66 allows the designer to analyse in detail these often complex relationships. When a component contributes to more than one function, it may be difficult to break down its overall cost into precise part costs per function. Approximate but well-informed estimates then have to be made.

(a)

Parts	Stop air	Sense ram air	Sense servo air	Sense cabin air	Connect parts	Provide mounting	Resist corrosion	Provide support	Provide interchangeability	No function	£ total cost	%
Banjo assembly			0.2		0.4				0.47		1.07	5.5
Valve body	0.4	1.0			2.82	0.8	0.2	0.8		0.6	6.62	34.0
Spring										0.39	0.39	2.0
Diaphragm assembly	0.6	0.1	0.1	0.1	0.94		0.2	0.1			2.14	11.0
Cover			0.4		1.2	0.1	0.1	0.34	0.1		2.24	11.5
Lug									0.1		0.1	0.5
Nuts, bolts and washers					2.14		0.1			0.1	2.34	12.0
Assembly					4.58						4.58	23.5
Total	1.0	1.1	0.7	0.1	12.08	0.9	0.6	1.24	0.67	1.09	19.48	100.0
% total	5.1	5.7	3.4	0.5	62.0	4.6	3.1	6.4	3.4	5.6		
High or low					H					H		

(a)

Parts	Stop air	Sense ram air	Sense servo air	Sense cabin air	Connect parts	Provide mounting	Resist corrosion	Provide seal	Provide interchangeability	Provide testing	£ total cost	%
Cover and connection	0.15	0.25	0.50	0.10	0.25	0.30		0.15	0.06		1.76	25.5
Body assembly	0.15	0.20	0.25	0.45	0.45	0.40		0.25	0.03		2.18	31.5
Diaphragm assembly	0.15	0.10	0.25	0.20	0.25	0.10		0.20	0.03		1.28	18.5
Valve assembly	0.05		0.05	0.05	0.15			0.31	0.05		0.66	9.5
Fasteners, nut bolts, etc.					1.04						1.04	15.0
Total	0.50	0.55	1.05	0.80	2.14	0.80		0.91	0.17		6.92	100.0
% total	7.2	7.9	15.1	11.6	30.9	11.6		13.2	2.5			
High or low					H							

(b)

Figure 66. (a) Function/cost analysis matrix for the air valve; (b) Function/cost analysis matrix for the redesigned air valve

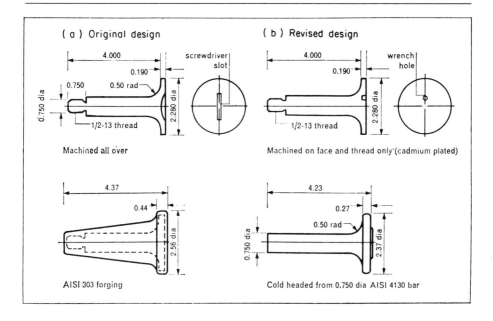

Cost reduction on the pintle			
	Cost of raw form	Cost of labour	Total cost ($)
Original	$2.31	$5.03	7.34
New	$0.59	$1.63	2.22
Gross saving/piece			5.12
Cost of implementation			1100
Savings on 5000 pieces			24 500

Figure 67. Redesign of a jet engine component following a value analysis

The example shown in Figure 66 is the analysis of an aircraft air valve. The analysis revealed the relatively high cost of the 'connect parts' function, as well as the redundancy of some elements. A redesign enabled some substantial reductions to be made, with a total cost saving of over 60 per cent. (Source: EITB.)

Example 4: Pintle

There have also been examples of significant cost savings resulting from the application of value analysis to relatively low-cost components. In this example, a high percentage cost reduction achieved by redesign of a simple, low-cost component resulted in major savings because very high numbers of the component were manufactured. The example is that of a pintle used as a flame holder and turbulence-increasing device in a jet engine. The size and shape of the component were fixed by all the other components around it; the value analysis exercise therefore concentrated on the material used and the manufacturing method. One of the detailed modifications made was to change from a screwdriver slot for assembling the pintle into the engine to an offset wrench hole; the slot needed an extra milling process in manufacture, whereas the hole could be made by a drilling head attached to the turret of the lathe.

Figure 67 shows the original and revised designs, and a costs table showing the approximately 70 per cent cost reduction per piece and the considerable overall savings on the large total number of pieces. (Source: Jones, 1981.)

Example 5: Piston

The elimination of unnecessary parts can be a significant factor in reducing the overall cost of an assembly, and is a principal focus of value analysis. Figure 68 shows the redesign of a small piston assembly, eliminating or combining several parts that were in the original. Separate fasteners should be eliminated wherever possible, and it was found in this example that the two screws could be eliminated

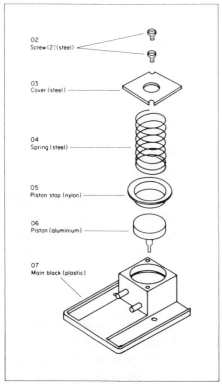

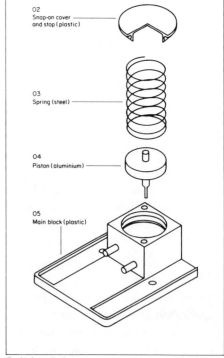

Original design for piston assembly   Redesigned piston assembly

Figure 68. Redesign of a piston assembly to reduce the number of components

by changing the cover plate from steel to plastic with a snap-fit onto the main block. The cover was also redesigned to incorporate the piston stop in a one-piece item. In the redesign, the number of parts was thus almost halved, resulting in reduced material and assembly costs with no loss of performance and an aesthetically improved product. (Source: Redford, 1983.)

Example 6: Electric motor

One of the main emphases that often emerges from a value engineering study is that of reducing the assembly time for a product, or of making the assembly

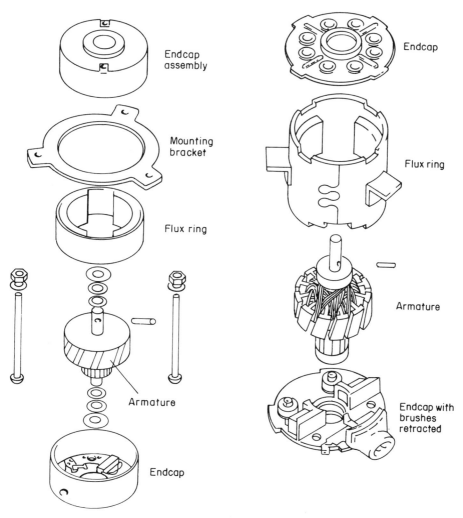

Figure 69. Redesign of a d.c. electric motor to reduce the number of components and assembly time

procedure suitable for automation. In this example, a study of the assembly process for an electric motor revealed that operators had to invert the part-assembled motor several times. A redesign enabled the procedure to be simplified so that the motor is built up in a sequential stacked assembly which does not need to be reorientated. The mounting bolts, nuts and washers of the original design were eliminated and replaced with a process in which the flux ring is crimped onto the endcaps. The original design had 56 parts and took more than seven minutes to assemble by hand; the redesign has 18 parts and can be assembled manually in less than three minutes, but will eventually be assembled automatically. (Source: ElMaraghy *et al.*, 1983.)

Example 7: Tubular heater

In a company manufacturing various kinds of electrical appliances, its range of tubular heaters was selected for a value engineering exercise. These heaters are simple and robust and used mainly in industrial and office premises to provide background heat. The product range consists of similar tubes of various lengths, providing various heat outputs at a standard wattage per unit length.

A component/function/cost analysis, shown in the chart below, revealed that the largest parts and labour cost was accounted for by what was regarded as the third most important function—that of providing the power connection. A closer examination of this function revealed two distinct sub-functions: firstly, providing

FUNCTION COST ANALYSIS FORM

Basic function: DISSIPATE HEAT

Item	Support functions	1 ENCLOSE ELEMENT	2 GENERATE HEAT	3 CONNECT POWER	5 PROVIDE APPEARANCE	4 PROTECT AGAINST DAMAGE	Cost of item P	%
Tube assy.		24·1					24·1	
Element assy.		6·8	24·8	0·6			32·2	
Term. collar assy.				14·9			14·9	
Cover assy.				4·1			4·1	
Final assy.		0·5	0·5				1·0	
Packing						5·4	5·4	
Painting					4·0		4·0	
Interconnector				13·8			13·8	
Cost of function		31·4	25·3	33·4	4·0	5·4	99·5	
% of total cost		31	25	34	4	5	≈100%	
Cost sequence		②	③	①	⑤	④		

Figure 70. Function/cost analysis of the tubular heater

REPORT SUMMARY FORM

Project ..... *Heat tube 5ft.* ..... Date ....

Item ..... *Tube interconnector* .....

Present	Proposed

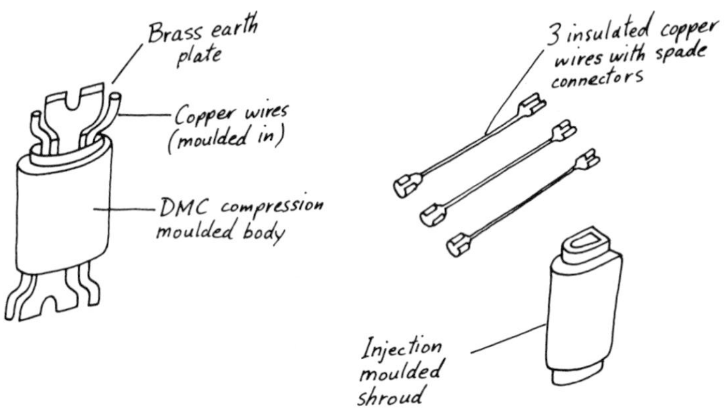

Brass earth plate

Copper wires (moulded in)

DMC compression moulded body

3 insulated copper wires with spade connectors

Injection moulded shroud

Comments

This proposal must be phased in at the same time as the termination proposal shown later. Tests have shown the proposal to be easier to install than the present method and there is an improved appearance

Costs	Material	Labour	Additional benefits	Total	
Present	13·0 p	0·8 p		13·8 p	
Proposed	4·0 p	—	*Optional item*	4·0 p	
Difference / % Saving	9·0 p / 69%	0·8 p / 100%		9·8 p / 71%	

Implementation time ..... *4 months* ..... Cost £ *1000* .....

Quantity per year ..... *70,000* ..... Expected life ..... *5 yrs.* .....

1st year savings £ ..... *~6000* ..... Annual savings £ ..... *~7000* .....

Figure 71. Proposals for redesign of the tube interconnector

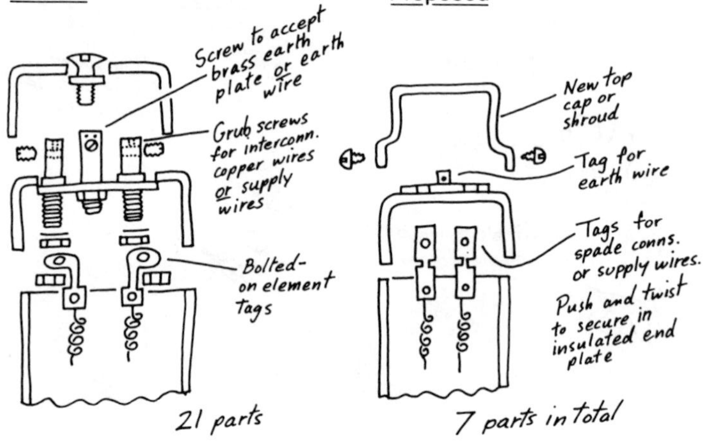

Figure 72. Proposals for redesign of the terminal cover assembly

an interconnector to allow tubes to be banked together on one mains connection and, secondly, providing a complex terminal connection.

An ideas-generating session produced suggestions for redesign which are shown in Figures 70 to 72. The moulded interconnector was replaced with three separate wires and a cover piece (Figure 71), which also enabled the terminal itself to be considerably simplified (Figure 72). Together, the modifications resulted in a cost reduction of 21 per cent.

Worked example: Handtorch

This relatively simple example demonstrates the principles both of applying value analysis with the objective of reducing a product's cost, and of applying value engineering with the objective of generating a more highly valued, innovative product.

Figure 73 shows how both value analysis and value engineering projects might start—with an exploded diagram of the product, which in this case is a conventional handtorch. The diagram shows the separate components and indicates how they are assembled together in the complete product.

Market research showed that two main aspects of a torch are highly valued by users. These are, firstly, the quality of the emitted light, perceived by users as being influenced by (apart from battery power) the bulb and the reflector, and, secondly, the ease of use of the torch, determined by the torch body and the switch. One low-valued feature of this particular torch design was the hanging loop on the base of the torch, which was hardly used at all and thus thought by most users to be redundant.

The components, their functions and perceived values are listed in Table 6, with values categorized simply as high, medium or low. It is useful to note that some components which may be important to the technical performance of the product are not necessarily perceived as of high value by users—examples here include the bulbholder and the pressure spring in the base.

A value analysis exercise led fairly quickly to some suggested modifications which would lower the product's cost without lowering its value. The reflector cover seemed to be too complicated, with its three separate components—glass, washer and screw-on retainer. A one-piece, clear plastic cap was suggested as an alternative. The base of the torch also seemed to be a rather complicated assembly, and again a one-piece plastic screw-on cap was suggested, with an integral plastic tongue spring to provide the pressure on the batteries, and the hanging loop eliminated. A proposal was also made to eliminate the switch, with electrical interrupt being provided instead by twisting the head of the torch. However, on evaluation it was decided that this was not very convenient for the user, and risked losing the highly valued ease of use of the thumb-switch.

Table 6 shows that the costed redesign indicated a potential saving in manufacturing costs of approximately 20 per cent.

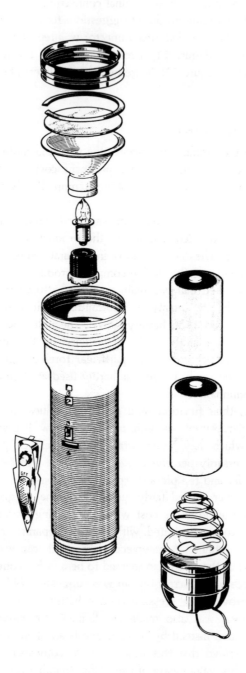

Figure 73. Exploded diagram of a handtorch

A more comprehensive value engineering exercise would have concentrated on the high-value aspects of the torch as perceived by users, and would have sought to improve these features, to enhance them or to generate innovations related to them. The high-value features of the torch are to do with its light beam and its handling.

Some research with users might well have found that the conventional torch has some shortcomings in these areas. For instance, it seems basically designed to throw a moderately wide beam over a fairly large distance—such as for illuminating a footpath. However, most use of a torch these days is for closer

TABLE 6

Component	Function	Value	Cost (£) Original	Redesign
Cap Washer Glass	Protect bulb and reflector	Medium	0.16	0.08
Reflector	Project light beam	High	0.12	0.12
Bulb	Provide light	High	0.10	0.10
Bulb holder	Hold bulb, provide electrical contact	Low	0.05	0.05
Torch body	Contain batteries, locate parts, provide hand grip	High	0.26	0.26
Switch	Provide electrical interrupt	High	0.08	0.08
Spring Washer	Provide pressure on batteries	Low	0.10	0.10
Cap	Protect batteries	Medium	0.10	
Loop	Provide for hanging	Low	0.03	—
Total			£1.00	£0.79

illumination such as finding a keyhole or making emergency repairs on a motorcar engine. In the latter type of case, it is important to be able to place the torch down, leaving one's hands free, and to direct the beam to the appropriate spot. The conventional, cylindrical torch is poorly designed for this; it is also inconveniently shaped for carrying in a pocket or handbag.

The novel 'Durabeam' torch (Figure 74) illustrates how these principles might have been applied in the design of a new product. The batteries are placed side by side instead of end to end, creating a flat, rectangular, compact body shape. The thumb-switch has been eliminated by using a 'flip-top' mechanism which acts as a switch and which also allows the angle of direction of the beam to be adjusted.

Figure 74. The Durabeam handtorch

Chapter 10

DESIGN STRATEGIES

What is a Design Strategy?

Using any particular design method during the design process will often appear to be diverting effort from the central task of designing. However, this is exactly the importance of using such a method—it involves applying some thought to the *way* in which the problem is being tackled. It requires some strategic thinking about the design process.

A design strategy describes the general plan of action for a design project and the sequence of particular activities (i.e. the tactics, or design methods) which the designer or design team expect to take to carry through the plan. To have a strategy is to be aware of where you are going and how you intend to get there. The purpose of having a strategy is to ensure that activities remain realistic with respect to the constraints of time, resources, etc., within which the design team has to work.

Many designers seem to operate with no explicit design strategy. However, having no apparent plan of action can be a strategy, of sorts! It might be called a 'random search' strategy, and might very well be appropriate in novel design situations of great uncertainty, where the widest possible search for solutions is being made. Examples of such novel situations might be trying to find applications for a completely new material, or designing a completely new machine such as a domestic robot.

For these kinds of situations, an appropriate strategy would be to search (at least to begin with) as widely as possible, hoping to find or generate some really novel and good ideas. The relevant tactics would be drawn mainly from the creative methods.

At the opposite extreme to 'random search' would be a completely predictable or 'prefabricated' sequence of well-tried-and-tested actions. Such a strategy would be appropriate in familiar and well-known situations. Again, it might not seem to be an explicit strategy, simply because it involves following a well-worn path of conventional activities. Examples of appropriate situations for such a strategy might include designing another variation of the machine that the designer's employer always makes or designing a specific and conventional type of product for an identified sector of the market.

In such situations, the design strategy would be aimed at narrowing the search for solutions and quickly homing-in on a satisfactory design. Relevant tactics would be drawn from conventional techniques and the rational methods.

Strategy styles

The 'random search' and 'prefabricated' strategies represent two extreme forms. In practice, most design projects require a strategy that lies somewhere between the two extremes and contains elements of both.

The 'random search' strategy represents a predominantly *divergent* design approach; the 'prefabricated' strategy represents a predominantly *convergent* approach. Normally, the overall aim of a design strategy will be to converge onto a final, evaluated and detailed design proposal. However, within the process of reaching that final design there will be times when it will be appropriate and necessary to diverge—to widen the search or to seek new ideas and starting points. The overall design process is therefore convergent, but it will contain periods of deliberate divergence (Figure 75).

Psychologists have suggested that some people are more naturally convergent thinkers and some are more naturally divergent thinkers. These preferred 'thinking styles' mean that some designers may be happier with one kind of strategy style rather than with another, one person may prefer a more convergent style whereas another may prefer a more divergent style. Alternatively, in a team context, designers with one preferred style may come to the fore in certain stages of the design process, and others come to the fore at other stages.

Convergent thinkers are usually good at detail design, at evaluation and at selecting the most appropriate or feasible proposal from a range of options. Divergent thinkers are usually good at concept design and at the generation of a wide range of alternatives. Clearly both kinds of thinking are necessary for

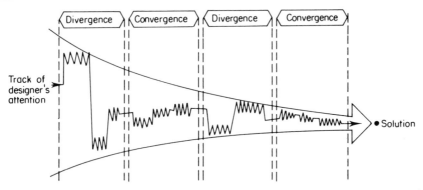

Figure 75. The overall process is convergent, but it includes periods of both convergence and
divergence

successful design. Unfortunately, much engineering (and other) education tends to promote and develop only convergent thinking.

As well as convergent and divergent, other kinds of 'thinking styles' have also been identified by psychologists, and may also have importance in design and in the structuring of design strategies. One of the most important dichotomies in thinking style appears to be that between *serialist* and *holist*. A serialist thinker prefers to proceed in small, logical steps, tries to get every point clear or decision made before moving on to the next, and pursues a straight path through the task, trying to avoid any digressions. A holistic thinker prefers to proceed on a much broader front, picking up and using bits of information that are not necessarily connected logically, and often doing things 'out of sequence'.

Another distinction that has been made between styles of thinking is that between *linear* and *lateral* thinking. Linear thinking proceeds quickly and efficiently towards a perceived goal, but may result in getting 'stuck in a rut' whilst lateral thinking entails a readiness to see, and to move to, new directions of thought.

The dichotomies of thinking style suggested by the psychologists tend to fall into two groups:

Convergent	Divergent
Serialist	Holist
Linear	Lateral

There is even some evidence to suggest that there is a fundamental dichotomy between the 'thinking styles' of the two hemispheres of the human brain. The left hemisphere predominates in rational, verbal, analytic modes of thought, whilst the right hemisphere predominates in intuitive, nonverbal, synthetic modes of thought.

Differences of thinking styles therefore appear to be an inherent characteristic of human beings. Most people *tend* towards a preference for one style rather than another, but no-one is *exclusively* limited to just one style. In particular, it is actually important to be able to change from one style to another in the course of a design project.

However, many models of the design process, such as those discussed in Chapter 2, do tend to present design as a linear, serialistic process. This may be off-putting and even counter productive to those designers whose own preferred thinking style tends more towards the lateral and holistic. What is needed is a more flexible, strategic approach to designing, which identifies and fosters the right kind of thinking at the right time and within the context of the particular design project.

Strategy analogies

To convey this more flexible approach to design strategies and tactics, some authors have resorted to the use of analogies. For example, J.C. Jones (1981) has suggested that a designer is like an explorer searching for buried treasure:

> A new problem is like an unknown land, of unknown extent, in which the explorer searches by making a network of journeys. He has to invent this network, either before he starts or as he proceeds.
>
> Design methods are like navigational tools, used to plot the course of a journey and maintain control over where he goes. Designing, like navigation, would be straightforward if one did not have to depend on inadequate information in the first place. Unlike the explorer's, the designer's landscape is unstable and imaginary, it changes form according to the assumptions he makes.
>
> The designer has to make as much sense as he can of every fragmentary clue, so that he can arrive at the treasure without spending a lifetime on the search. Unless he is very unlucky, or very stupid, he will come across the treasure long before he has searched every inch of the ground.

Koberg and Bagnall (1974) have suggested that the designer is like a traveller and that 'the design process is a problem-solving journey':

> A general rule is to find and use those methods which best fit the problem as well as the abilities of the problem solver. It's a task similar to that of selecting the route, side roads and overnight stops for an auto trip. Just as any competent trip planner would examine the alternative routes on a map, and read through several brochures, books or articles before choosing a route for his trip, so should the problem solver review the methods available, and not be afraid to adapt any of them to his special needs.

Instead of exploring or travelling, I prefer an analogy based on football. A design team, like a football team, has to have a strategy. The football team's strategy for defeating the opposition will consist of an agreed plan to use a variety of plays or moves (i.e. techniques or methods), to be applied as the situation demands. During the game, the choice of a move, and whether or not it is successful, will depend on the specific circumstances, on the skill of the players, and on the response of the opposition.

The repertoire of moves used in a game is partly decided in advance, partly improvised on the field, and also amended at the half-time briefing by the team manager or coach. The manager's role is important because he maintains a wider view of the game than the players can have out there on the field. In designing, it is necessary to adopt a similar role from time to time, in reviewing your project's strategy and progress.

For you as an individual designer, or member of a design team, tackling your problem and reaching your goal will involve both the strategic skills of the manager and the tactical skills of the player. Also, like the team, you will have to make on-field and half-time reviews of your strategy to ensure that your problem does not defeat you!

A design strategy, therefore, should provide you with two things:
1. A *framework* of intended actions within which to operate,
2. A management *control* function enabling you to adapt your actions as you learn more about the problem and its 'responses' to your actions.

Frameworks for Action

One 'framework', complete with appropriate methods identified and located within it, has already been suggested. That was the model of the design process outlined in Chapter 3:

Stages in the design process	*Appropriate methods*
1. Clarifying objectives	Objectives tree
2. Establishing functions	Function analysis
3. Setting requirements	Performance specification
4. Generating alternatives	Morphological chart
5. Evaluating alternatives	Weighted objectives
6. Improving details	Value engineering

If it seemed to be appropriate to the specific project in hand, then you could adopt this as a complete 'prefabricated' strategy. It comprises a six-stage framework covering the design process from customer requirements through to detail design and a suitable 'tactic'—a design method—for each stage. You could, of course, add, or substitute, methods in each stage. For example, you could use brainstorming instead of a morphological chart as a way of generating alternative solutions; you could use the conventional 'design-by-drawing' method instead of, or perhaps as well as, value engineering or analysis at the stage of detail design.

However, this particular framework does imply that the design process is going to be a fairly straightforward, step-by-step process. It implies a 'linear' design process. A design strategy that is more suited to a 'lateral' approach might be something like this:

Stage	Tactics to be used
1. Divergent problem exploration	Morphological chart
	Brainstorming
2. Structuring of problem	Objectives tree
	Performance specification
3. Convergence on solution	Synectics

Another framework might be adopted from the general pattern of the creative process, as outlined in Chapter 3. This might be developed as follows:

Stage	Tactics to be used
1. Recognition	Brainstorming
	Writing a design brief
2. Preparation	Objectives tree
	Information search
	Function analysis
3. Incubation	Taking a holiday
	Talking the problem over with colleagues and friends
	Tackling another problem
	Enlarging the search space: counterplanning
4. Illumination	Morphological chart
	Brainstorming
	Enlarging the search space: random input
5. Verification	Performance specification
	Weighted objectives

You see therefore that there can be many different strategy frameworks, and many different tactical combinations of methods and techniques.

Strategy Control

The second important aspect of a successful design strategy is that it has a strong element of management control built into it. If you are working alone on a project, then this means self-management, of course. If you are working in a team then either the team leader or the whole team, collectively, must from time to time review progress and amend the strategy and tactics if necessary.

Whatever general framework is adopted for the project, it is necessary to have some further strategy control in order to avoid unnecessary time-wasting, pursuing red herrings and the like. Some simple rules of strategy control are:

1. Keep your objectives clear. In designing, it is impossible to have one set of completely fixed objectives, because ends and means are inextricably interwoven in the product you are designing. A creative resolution of a design problem often involves changing some of the original objectives. However, this does not mean that it is impossible to have any clear objectives at all. On the contrary, it is important to have your objectives clear at any time (probably in the form of an 'objectives tree'), but also to recognize that they can change as your project evolves.

2. Keep your strategy under review. Remember that your overall aim is to solve the design problem in a creative and appropriate way—it is not doggedly to follow a path you have set for yourself that might be leading nowhere! A design strategy needs to be flexible, adaptable and intelligent, so review it regularly. If you feel that your actions are not being very productive, or that you are getting stuck, then pause to ask yourself if there is a better way of proceeding. Have confidence in adapting the 'tactics'—the methods and techniques—to your own ways of working and to the aims and progress of the project.

3. Involve other people. Different people 'see' a problem in different ways, and it is often true that 'two heads are better than one'. If you are getting stuck, one of the easiest ways to sort out what is going wrong is to explain the project to someone else—a colleague or a friend. Other people, of course, are also able to offer ideas and different viewpoints on the problem which may well suggest ways to change your approach.

4. Keep separate files for different facets. There will, almost certainly, be times when you are having to work on several different facets of a project in parallel, so keep separate files which allow you to switch rapidly from one facet to another or to take in a new piece of information in one area without distracting your work on another. One very useful file to keep is 'solution ideas'. You will probably come across, or have ideas for solutions at all times throughout the project, but you will need just to keep them filed until you are ready to turn your whole attention to solution concepts or details.

Exercises: Choosing Strategies and Tactics

These short exercises are intended to give some practice in devising strategic 'frameworks' and selecting appropriate tactical methods or techniques. Each exercise need only take 5–10 minutes.

Exercise 1

Your company manufactures industrial doors of various kinds. With the increased availability of electronic devices, remote controls, and so on, the company has decided to produce a new range of automatically operated doors. You have been asked to propose a set of prototype designs that will establish the basic features of this new range. Outline your design strategy and tactics.

Exercise 2

Your company manufactures packing machinery. One of the company's most valued customers is about to change its product range and will therefore need to replace its packing machinery. You will be responsible for designing this new machinery. Outline your design strategy and tactics.

Exercise 3

You have just been appointed design consultant to a company manufacturing office equipment. Its sales have fallen drastically because its designs have failed to keep up with modern office equipment trends. To reestablish its position the company wants a completely new product that will be a step ahead of all its rivals. You have to suggest what the new product should be and produce some preliminary design proposals for a Board meeting in two weeks' time. Outline your design strategy and tactics.

Discussion of exercises

1. *Industrial doors.* The change from manual to automatic doors implies that there could be scope for rethinking the scope of the company's current range— perhaps to include some door types that were not previously included. It is therefore worth putting some divergent search effort into the early stages of the design project. My own suggested strategy would be:

Framework	Tactics
1. Problem exploration	Brainstorming
	Synectics
2. Problem specification	Function analysis
3. Alternative solutions	Morphological chart
4. Selection of alternatives	Weighted objectives

2. *Packing machinery.* This appears to be straightforward case of redesigning an established product. My suggested strategy would be:

Framework	Tactics
1. Customer requirements/ Problem specification	Performance specification
2. Alternative solutions/ Evaluation of alternatives	Value engineering
3. Detail design	Conventional design by drawing

3. *Office equipment.* This problem suggests the need for some radical design thinking—pretty quickly! My suggested strategy would be:

Framework	Tactics
1. Divergent search	Enlarging the search space: why-why-why? Function analysis
2. Alternative solutions	Brainstorming Morphological chart
3. Convergent selection	Objectives tree

I think that I might try to use a modified conversion of the 'objectives tree' method in the final stage, and to work this into my presentation to the Board so as to relate the choice of alternatives to the company's objectives.

REFERENCES AND SOURCES

Archer, L.B. (1984).
Systematic method for designers.
In N. Cross (ed.), *Developments in Design Methodology*,
Wiley, Chichester.

Cross, N. (1984).
Developments in Design Methodology,
Wiley, Chichester.

Ehrlenspiel, K. (1987).
Reduction of Product Costs.
International Conference on Engineering Design,
American Society of Mechanical Engineers, New York.

Ehrlenspiel, K., and T. John (1987).
Inventing by Design Methodology.
International Conference on Engineering Design,
American Society of Mechanical Engineers, New York.

153

ElMaraghy, H.A., *et al.* (1988).
Redesign for assembly of a direct current motor.
Developments in Assembly Automation,
IFS Ltd., Bedford.

Engineering Industry Training Board (no date).
Value Engineering,
EITB, London.

French, M.J. (1985).
Conceptual Design for Engineers,
Design Council, London.

Hawkes, B., and R. Abinett (1984).
The Engineering Design Process,
Pitman, London.

Hubka, V. (1982).
Principles of Engineering Design,
Butterworth, London.

Jones, J.C. (1981).
Design Methods,
Wiley, Chichester.

Jones, J.C. (1984).
A method of systematic design.
In N. Cross (ed.), *Developments in Design Methodology,*
Wiley, Chichester.

Koberg, D., and J. Bagnall (1974).
The Universal Traveler,
Kaufmann, Los Altos, Calif.

Krick, E. (1976).
An Introduction to Engineering,
Wiley, New York.

Lawson, B.R. (1984).
Cognitive strategies in architectural design.
In N. Cross (ed.), *Developments in Design Methodology,*
Wiley, Chichester.

Love, S.F. (1980).
Planning and Creating Successful Engineered Designs,
Advanced Professional Development, Los Angeles, Calif.

Luckman, J. (1984).
An approach to the management of design.
In N. Cross (ed.), *Developments in Design Methodology*,
Wiley, Chichester.

March, L.J. (1984).
The logic of design.
In N. Cross (ed.), *Developments in Design Methodology*,
Wiley, Chichester.

Marples, D. (1960).
The Decisions of Engineering Design,
Institute of Engineering Designers, London.

Norris, K.W. (1963).
The morphological approach to engineering design.
In J. C. Jones and D. Thornley (eds.), *Conference on Design Methods*,
Pergamon, Oxford.

Pahl, G., and W. Beitz (1984).
Engineering Design,
Design Council, London.

Pighini, U., *et al.* (1983).
The determination of optimal dimensions for a city car,
Design Studies, **4**, No. 4.

Pitts, G. (1973).
Techniques in Engineering Design,
Newnes-Butterworths, London.

Redford, A.H. (1983).
Design for assembly,
Design Studies, **4**, No. 3.

Shahin, M.M.A. (1988).
Application of a systematic design methodology,
Design Studies, **9**, No. 4.

Simon, H.A. (1984).
The structure of ill-structured problems.
In N. Cross (ed.), *Developments in Design Methodology*,
Wiley, Chichester.

Tebay, R., J. Atherton and S.H. Wearne (1984).
Mechanical engineering design decisions,
Proc. Instn. Mech. Engrs., **198B**, No. 6.

Tjalve, E. (1979).
A Short Course in Industrial Design,
Newnes-Butterworths, London.

Tovey, M. (1986).
Thinking styles and modelling systems,
Design Studies, **7**, No. 1.

INDEX